Nonlinear Systems and Complexity

Volume 39

Series Editor

Albert C. J. Luo, Southern Illinois University, Edwardsville, IL, USA

Nonlinear Systems and Complexity provides a place to systematically summarize recent developments, applications, and overall advance in all aspects of nonlinearity, chaos, and complexity as part of the established research literature, beyond the novel and recent findings published in primary journals. The aims of the book series are to publish theories and techniques in nonlinear systems and complexity; stimulate more research interest on nonlinearity, synchronization, and complexity in nonlinear science; and fast-scatter the new knowledge to scientists, engineers, and students in the corresponding fields. Books in this series will focus on the recent developments, findings and progress on theories, principles, methodology, computational techniques in nonlinear systems and mathematics with engineering applications. The Series establishes highly relevant monographs on wide ranging topics covering fundamental advances and new applications in the field. Topical areas include, but are not limited to: Nonlinear dynamics Complexity, nonlinearity, and chaos Computational methods for nonlinear systems Stability, bifurcation, chaos and fractals in engineering Nonlinear chemical and biological phenomena Fractional dynamics and applications Discontinuity, synchronization and control.

Fuhong Min

Discontinuous Dynamics and System Synchronization

Springer

Fuhong Min
Nanjing Normal University
Nanjing, China

ISSN 2195-9994 ISSN 2196-0003 (electronic)
Nonlinear Systems and Complexity
ISBN 978-3-031-66647-6 ISBN 978-3-031-66648-3 (eBook)
https://doi.org/10.1007/978-3-031-66648-3

This Springer imprint is published by the registered company Springer Nature Switzerland AG
The registered company address is: Gewerbestrasse 11, 6330 Cham, Switzerland

Preface

The history of the study of discontinuous dynamical systems can be traced back to the 1930s. In 1931, Hartog studied the periodic motion of damped linear oscillators under friction and forcing. In 1964, Filippov established the theory of nonsmooth systems by introducing the concepts of "differential inclusion" and "extremal function." However, in 2005, Luo introduced the G-function and subsequently proposed the theory of discontinuous dynamical systems on time-varying domain, which led to a new wave of research on discontinuous dynamical systems.

This book focuses on the dynamical behavior of discontinuous dynamical systems at the boundary. The book is divided into two parts: The first part deals with the flow switching theory of discontinuous dynamical systems, which is partially covered in Chaps. 2–5. The second part is about discontinuity theory of the interaction between two dynamical systems, which is partially covered in Chaps. 6–9. Chapter 1 discusses the existence and research significance of discontinuous dynamical systems. Chapter 2 describes the flow switching theory of discontinuous dynamical systems. Chapter 3 investigates the nonlinear dynamics of Duffing-like systems with signum function. The non-autonomous FitzHugh-Nagumo neuron circuit containing piecewise memristor is investigated in Chap. 4. Utilizing the theory of discontinuous dynamical systems, the complex dynamics of non-smooth Hindmarsh–Rose (HR) neuron system with locally active memristor is investigated in Chap. 5. Chapter 6 describes the discontinuous theory of the interaction of two dynamical systems. In Chap. 7, the partial and complete chaotic synchronizations of two nonlinear gyroscope systems in the presence and absence of noise are investigated. The generation mechanism of the projective synchronization of Chua circuits with different scrolls is thoroughly investigated in Chap. 8. Chapter 9 discusses the interactions of a van der Pol system, a Pendulum system and a Duffing oscillator under a switching nonlinear controller with symbolic functions. This book may not contain all the theory of discontinuous dynamical systems, and there may even be some inaccurate descriptions. The reader is kindly requested to make valuable suggestions for corrections.

Finally, I would like to thank my students (Jiayun Chen, Hanyun Ma, Wen Zhang, Chen Wei, and Zhi rui) for completing all numerical computations and circuit implementations. I also thank my advisor Albert C.J. Luo for his help and guidance. Herein, I thank my husband (Caojun Ge) and my daughter (Yixiao Ge) for their tolerance, patience, understanding, and support.

Nanjing, China
March, 2024

Fuhong Min

Contents

1 Introduction . . . 1
1.1 Discontinuous Systems . . . 2
1.2 Book Layout . . . 4
References . . . 5

2 The Flow Passability Theory . . . 9
2.1 Basic Concepts . . . 9
2.2 The Flow Passability Theory . . . 13
References . . . 21

3 A Duffing-Like Oscillator with Signum Nonlinearity . . . 23
3.1 Discontinuous Dynamical System . . . 23
3.2 Analytical Switching Conditions . . . 25
3.3 Mapping Structures . . . 27
3.4 Bifurcation Scenarios . . . 29
3.5 Parameter Maps . . . 31
3.6 Simulation Illustrations . . . 32
References . . . 34

4 Switching Dynamics of a FitzHugh-Nagumo Circuit . . . 37
4.1 Memristor-Based FitzHugh-Nagumo Circuit . . . 37
4.2 Analytical Conditions for Motion Switching . . . 40
4.3 Generic Mappings . . . 43
4.4 Bifurcation Scenarios . . . 45
4.5 Distribution Maps . . . 50
4.6 Switching Dynamics . . . 52
References . . . 56

5 A Memristive Hindmarsh-Rose Neuron System . . . 57
5.1 Memristor-Based Hindmarsh-Rose Circuit . . . 57
5.2 Switchability Conditions . . . 62
5.3 Switching Sets and Mappings . . . 64

5.4 Bifurcation Scenarios . . . 65
5.5 Firing Phenomenon . . . 68
5.6 Discontinuous Dynamics . . . 68
5.7 Hardware Implementation . . . 73
References . . . 75

6 Dynamical System Interaction Theory . . . 77
6.1 Introduction to System Interactions . . . 77
6.2 An Interaction Theory . . . 82
6.3 Edge Interaction Theory . . . 88
References . . . 91

7 Two-Identical System Synchronization . . . 93
7.1 Introduction to Gyroscope System Synchronization . . . 93
7.2 Discontinuous Dynamics . . . 95
7.3 Synchronization and Attractivity . . . 104
7.4 Parameter Characteristics . . . 107
7.5 Numerical Simulations . . . 110
References . . . 112

8 Chua System Synchronization and Implementation . . . 113
8.1 Slave and Master Systems . . . 113
8.2 Absolute Description . . . 114
8.3 Relative Description . . . 120
8.4 Analytical Synchronization Dynamics . . . 126
8.5 Illustrations of Synchronization Dynamics . . . 134
8.5.1 Synchronization Scenarios . . . 135
8.5.2 Parameter Maps . . . 136
8.5.3 Synchronization Illustrations . . . 137
8.6 Circuit Implementation . . . 142
References . . . 149

9 Function Synchronizations of Dynamic Systems . . . 151
9.1 Description of Systems . . . 151
9.2 Switching Synchronization Dynamics . . . 155
9.3 Switching Projective . . . 159
9.4 Synchronization Scenario and Parameter Maps . . . 163
9.5 Numerical Illustration . . . 164
References . . . 170

Index . . . 171

Chapter 1
Introduction

Contents

1.1 Discontinuous Systems 2
1.2 Book Layout 4
References 5

In the field of engineering, discontinuous dynamical systems are ubiquitous. However, people often tend to use continuous system models to approximate and explore the dynamical characteristics of discontinuous systems for the sake of convenience. Nevertheless, continuous system models often struggle to accurately depict discontinuous systems, potentially leading to problems becoming ambiguous and more complex. This is because discontinuities are omnipresent in discontinuous systems. Therefore, to more precisely describe real-world systems in engineering and perform theoretical analysis and practical predictions, we must employ discontinuous models to characterize these discontinuous systems. Discontinuous dynamical systems can be viewed as composed of numerous continuous subsystems that operate in different domains or at different time intervals. However, the dynamical properties of these subsystems differ significantly from those of interconnected continuous subsystems. Hence, discontinuous dynamical systems can primarily be classified into two categories: the first involves systems with different dynamics in two adjacent time intervals. When one dynamical system reaches a switching point, the system automatically transitions to another dynamical system. The second category pertains to situations where the dynamical systems are distinct in any two neighboring domains. When the flow of subsystems reaches a boundary, it is necessary to provide transmission theorems or transition rules at the boundary. In this book, we will predominantly focus on introducing the fundamental theory of discontinuous dynamical systems. And we will apply this theory to explore the interaction between two dynamical systems. Finally, this chapter will provide an overview of the book's structure and the content of each chapter.

F. Min, *Discontinuous Dynamics and System Synchronization*, Nonlinear Systems and Complexity 39, https://doi.org/10.1007/978-3-031-66648-3_1

1.1 Discontinuous Systems

Most of the existing theories in dynamical systems are based on the Lipschitz condition (e.g., Poincaré, 1892; Birkhoff, 1927), which are extensively adopted in science and engineering. However, the Lipschitz condition is very strong for practical dynamical problems, and not adequate for the discontinuous dynamical systems. Early research on discontinuous dynamical systems originated in the field of mechanical engineering, where discontinuities were primarily caused by frictional forces (e.g., Den Hartog, 1930, 1931; Den Hartog & Mikina, 1932). Levinson (1949) studied the van der Pol equation with periodic excitation using a piecewise linear model and discovered an infinite number of undisturbed periodic motion solutions. Levitan (1960) investigated friction-induced oscillators driven by periodic external forces and conducted an analysis of periodic motion. Filippov (1964) examined the motion of a Coulomb friction oscillator and introduced a differential equation theory based on right-hand discontinuity. To study sliding motion on discontinuous boundaries, the concept of differential inclusions was introduced, and the existence and uniqueness of solutions to discontinuous differential equations were discussed in Filippov (1988). Aizerman and Pyatnitskii (1974) expanded the Filippov's theory and summarized the general theory of discontinuous dynamical systems. From such a generalized theory, the methods for controlling dynamic systems through the sliding modes control were developed in Utkin (1977, 1978, 1981) and DeCarlo et al. (1988), and the Filippov theory was used to investigate bifurcations in nonlinear discontinuous systems by Leine et al. (2000). More discussion about the traditional analysis of bifurcation in non-smooth dynamical systems can be referred to Zhusubaliyev and Mosekilde (2003). Discontinuous dynamical systems are a type of dynamical model that describes the interactions of multiple objects from a dynamical perspective. These objects interact in the real world, rather than existing individually, statically, or in isolation. Many practical problems can be modeled as discontinuous dynamical systems because they typically involve interactions between multiple objects. In the past, continuous dynamical systems were often used to describe such scenarios, but this approach made the problems more complex and less accurate. Therefore, researchers introduced new theories to address the challenges posed by discontinuous dynamical systems. Luo (2005a, b, 2006, 2007) proposed a general theory for addressing singularities in discontinuous dynamical systems. To study the motion of sources and sinks in discontinuous dynamical systems, he introduced the concepts of virtual flows, sink flows, and source flows. Subsequently, Luo and Gegg (2006a, b) used the theory of local singularities to study dry friction linear oscillators and proposed the force-product criterion for velocity switching at the boundary. Luo and Gegg (2006c, d) developed the dynamics of a friction-induced oscillator contacting on time-varying belts with friction. Lu (2007) mathematically proved the existence of periodic motion for time-varying friction oscillators. Luo and Thapa (2009) introduced a new model for a brake system consisting of two oscillators connected by a frictional contact surface. They used this model to study the nonlinear dynamical behavior of the brake system under periodic excitation. Luo (2008) investigated the switching behavior of flows on discontinuous boundaries by introducing the concept of G-functions. Luo and Huang (2012) studied the flow switchability at the boundary for the nonlinear, friction-induced periodically forced oscillators and established analytic conditions for stick and nonstick motions in such an oscillator. From the analytic conditions, the switchability of the periodic and chaotic motions was discussed. Luo (2011) introduced the concept of flow barriers in discontinuous dynamical systems and presents the passability of flow to the separation boundary with flow barriers, provided a theoretic base to further develop control theory and stability, and the theory of discontinuous dynamical systems was proposed in Luo (2012). Huang and Luo (2018) investigated the analytical conditions for switching impact-alike chatter at corners from the local theory of flow at the corner in discontinuous dynamical systems.

The emergence of the theory of discontinuous dynamical systems is attributed to the relentless exploration of some outstanding researchers. This book will focus on the work of these notable researchers, summarizing their contributions and exploring the profound impact of the theory of discontinuous dynamical systems in various fields. Luo's research laid a solid foundation for the development of the theory of discontinuous dynamical systems. He applied the theory to several typical dynamical models and provided detailed analytical methods. Notably, based on the theory of discontinuous dynamical systems, he developed a synchronization theory applicable to dynamical systems with specific constraints. By introducing G-functions, he successfully determined the switchability of flows in discontinuous systems, enabling synchronization between various nonlinear systems using different feedback controllers. In further developments in this field, Luo and Min (2010) proposed analytical conditions for the synchronization of forced Duffing oscillators with chaotic pendulums, deriving conditions for synchronization invariant domains. Luo and Min (2011a) considered a chaotic pendulum as the master system and a periodically forced Duffing oscillator as the slave system, discussed the partial and full synchronization of the two systems, and gave the analytical conditions for the two-system synchronization. Luo and Min (2011b) introduced the conditions for synchronization, asynchronization, and instantaneous synchronization, establishing functional relationships between the motion trajectories of controlled Duffing systems and Pendulum systems. Through numerical simulations, he demonstrated the effectiveness of this method. Min and Luo (2015); Min and Ma (2020) separately applied the theory of discontinuous dynamical systems to the projection synchronization of Chua's circuits and the combined switching synchronization of three nonautonomous oscillators, investigating analytical conditions and parameter characteristics for chaotic synchronization. Additionally, Gao and Min (2021) researched on switching theory provided insights into the current-controlled DC-DC buck converter under different operating conditions, particularly highlighting the boundary motion of MOS transistor switching times across various frequency ranges. This research is of great significance in the field of power electronics. Min and Zhang (2021) also explored the FHN neuron circuit with a piecewise linear flux-controlled memristor, delving into the phenomena of extreme multistability and coexistence of attractors. This expanded the scope of application of the theory of discontinuous dynamical systems and helped us better understand the complex dynamical behavior of neuron circuits. Through this book, we will guide readers into the essence of the theory of discontinuous dynamical systems and explore its wide-ranging applications in various fields, providing profound theoretical insights and practical tools for solving complex dynamical problems in the real world.

1.2 Book Layout

The book primarily consists of the following chapters:

In Chap. 2, the convertibility theory of boundary flows in discontinuous systems is introduced. The concept of G-functions is presented and utilized for analyzing the singularities near the boundaries in these systems. Furthermore, the chapter discusses the convertibility of flows from one region to another based on G-functions. Concepts like real flows and virtual flows are detailed, and various flow types near the boundaries, such as confluence, semi-confluence, source flow, semi-source flow, and crossable flow, are examined. Finally, the chapter introduces the transition bifurcations of crossing and non-crossing flows on the boundaries of discontinuous systems.

In Chap. 3, a study is conducted on the nonlinear behavior of the Duffing-like oscillator system with signum function. Analytical conditions for switching motion are provided to explain the system's behavior. Basic mappings and switching sets are used to describe different motion patterns in this nonsmooth dynamical system. Subsequently, the chapter explores the system's dynamical behavior through switch bifurcation diagrams, trajectories in phase plane, parameter mappings, and basins of attractions. To validate the numerical results' accuracy, a hardware circuit for the Duffing-like system is implemented using field-programmable gate array (FPGA).

In Chap. 4, the switching dynamics of the nonautonomous FitzHugh-Nagumo neuronal system with a three segment piecewise linear memristor are investigated. The chapter begins by studying the analytical conditions for crossing and grazing motion on the boundary, providing mapping structures to explain various periodic orbits and chaotic motion resulting from boundary separations. The impact of initial conditions and system parameters on the dynamic behavior is explored through switching bifurcation diagrams, phase plots, basins of attractions, and distribution maps, illustrating extreme multistability, including the coexistence of periodic orbits and chaotic attractors. Boundary dynamical simulations validate the effects of pastoral and grazing conditions.

In Chap. 5, complex dynamics in the Hindmarsh-Rose neuron system with a locally active memristor are examined. The chapter begins by designing equivalent circuits for the locally active memristor and neuron model, offering analytical conditions for various types of motion at the boundary. G-functions are employed to understand the motion mechanism of the memristive Hindmarsh-Rose

(HR) neuron. Various technical tools, such as bifurcation diagrams, phase trajectories, two-parameter diagrams, and basins of attractions, are used to study the bursting, spiking, and extreme multistability of the memristive HR neuron. An equivalent neural morphological circuit is developed using a field-programmable gate array (FPGA) to validate the numerical results.

In Chap. 6, a general theory is presented for the interaction of two dynamical systems. Conditions for the interaction of two dynamical systems at separating boundaries are discussed, considering time-varying separating boundaries. The chapter utilizes the theory of discontinuous dynamical systems to analyze the necessary and sufficient conditions for the interaction of two dynamical systems and provides a general analytical framework for this analysis.

In Chap. 7, the partial chaos synchronization and complete chaos synchronization of two nonlinear gyroscope systems with/without noise are studied. The chapter first determines the analytical conditions for synchronization based on boundary flows. It introduces scenarios for transitioning between asynchronous and synchronous states and explores the parameter characteristics of synchronization to provide a better understanding of the synchronization mechanism in controlled noiseless gyroscopic systems. Control parameter maps are implemented for further analysis.

In Chap. 8, the analytical conditions and mechanism of chaotic motion projection synchronization of different scrolls in Chua's circuits are discussed. Analytical conditions for chaotic motion in Chua's circuits are derived, establishing parameter diagrams for the beginning and disappearance of projection synchronization. The chapter provides detailed explanations of partial and complete projection synchronization based on parameter mappings and invariant domains under different scrolls in Chua's circuits. Experimental implementation of complete projection synchronization is compared to numerical calculations.

In Chap. 9, three nonautomatic system interactions, including a van der Pol system, a Pendulum system and a Duffing oscillator system with a signum function, are studied under nonlinear switching control. The chapter discusses three forms of interactions in nonautonomous systems under switching control, employing the theory of discontinuous dynamical systems. It presents the mechanisms for switching function projection synchronization and the interaction of these three systems. Numerical evidence for switching function projection synchronization and the effectiveness of control parameters is provided, along with the design and development of analog circuits for validation.

References

Aizerman, M. A., & Pyatnitskii, E. S. (1974). Foundation of a theory of discontinuous systems. *Automatic and Remote Control, 35*, 1066–1079.

Birkhoff, C. D. (1927). On the periodic motions of dynamical systems. *Acta Mathematica, 50*, 359–379.

DeCarlo, R. A., Zak, S. H., & Matthews, G. P. (1988). Variable structure control of nonlinear multivariable systems: A tutorial. *Proceedings of the IEEE, 76*, 212–232.

Den Hartog, J. P. (1930). Forced vibration with combined viscous and Coulomb damping. *Philosophical Magazine, VII*(9), 801–817.

Den Hartog, J. P. (1931). Forced vibrations with Coulomb and viscous damping. *Transactions of the American Society of Mechanical Engineers, 53*, 107–115.

Den Hartog, J. P., & Mikina, S. J. (1932). Forced vibrations with non-linear spring constants. *ASME Journal of Appiled Mechanice, 58*, 157–164.

Filippov, A. F. (1964). Differential equations with discontinuous right-hand side. *American Mathematical Society Translations, Series, 2*(42), 199–231.

Filippov, A. F. (1988). *Differential equations with discontinuous righthand sides*. Kluwer Academic Publishers.

Gao, P. Y., & Min, F. H. (2021). Dynamical analysis of boundary behaviors of current-controlled DC–DC buck converter. *Nonlinear Dynamics, 106*, 2203–2228.

Huang, J. Z., & Luo, A. C. J. (2018). Complex dynamics of bouncing motions at boundaries and corners in a discontinuous dynamical system. *Journal of Computational and Nonlinear Dynamics, 12*, 061014 (11).

Leine, R. I., van Campen, D. H., & van de Vrande, B. L. (2000). Bifurcations in nonlinear discontinuous systems. *Nonlinear Dynamics, 23*, 105–164.

Levinson, N. (1949). A second order differential equation with singular solutions. *Annals of Mathematics, 50*, 127–153.

Levitan, E. S. (1960). Forced oscillation of a spring-mass system having combined coulomb and viscous damping. *Journal of the Acoustical Society of America, 32*, 1265–1269.

Lu, C. (2007). Existence of slip and stick periodic motions in a non-smooth dynamical system. *Chaos, Solitons and Fractals, 35*, 949–959.

Luo, A. C. J. (2005a). A theory for non-smooth dynamical systems on connectable domains. *Communication in Nonlinear Science and Numerical Simulation, 10*, 1–55.

Luo, A. C. J. (2005b). Imaginary, sink and source flows in the vicinity of the separatrix of non-smooth dynamic system. *Journal of Sound and Vibration, 285*, 443–456.

Luo, A. C. J. (2006). *Singularity and dynamics on discontinuous vector fields*. Elsevier.

Luo, A. C. J. (2007). On flow switching bifurcations in discontinuous dynamical system. *Communications in Nonlinear Science and Numerical Simulation, 12*, 100–116.

Luo, A. C. J. (2008). A theory for flow swtichability in discontinuous dynamical systems. *Nonlinear Analysis: Hybrid Systems, 2*, 1030–1061.

Luo, A. C. J. (2011). On flow barriers and switchability in discontinuous dynamical systems. *Nonlinear Analysis: Hybrid Systems International Journal of Bifurcation and Chaos, 21*, 1–76.

Luo, A. C. J. (2012). Discontinuous dynamics systems. In *Communication in nonlinear science and numerical simulation*. Higher Education Press/Springer.

Luo, A. C. J., & Gegg, B. C. (2006a). On the mechanism of stick and non-stick periodic motion in a forced oscillator including dry-friction. *ASME Journal of Vibration and Acoustics, 128*, 97–105.

Luo, A. C. J., & Gegg, B. C. (2006b). Stick and non-stick periodic motions in a periodically forced, linear oscillator with dry friction. *Journal of Sound and Vibration, 291*, 132–168.

Luo, A. C. J., & Gegg, B. C. (2006c). Periodic motions in a periodically forced oscillator moving on an oscillating belt with dry friction. *ASME Journal of Computational and Nonlinear Dynamics, 1*, 212–220.

Luo, A. C. J., & Gegg, B. C. (2006d). Dynamics of a periodically excited oscillator with dry friction on a sinusoidally time-varying, traveling surface. *International Journal of Bifurcation and Chaos, 16*, 3539–3566.

Luo, A. C. J., & Huang, J. Z. (2012). Discontinuous dynamics of a non-linear, self-excited, friction-induced, periodically forced oscillator. *Nonlinear Analysis-Real World Applications, 13*, 241–257.

Luo, A. C. J., & Min, F. H. (2010). The chaotic synchronization of a controlled pendulum with a periodically forced, damped Duffing oscillator. *Communications in Nonlinear Science and Numerical Simulation, 16*, 4704–4717.

Luo, A. C. J., & Min, F. H. (2011a). Synchronization of a periodically forced Duffing oscillator with a periodically excited pendulum. *Nonlinear Analysis-Real World Applications, 12*, 1810–1827.
Luo, A. C. J., & Min, F. H. (2011b). Synchronization dynamics of two different dynamical systems, *Communications in Nonlinear Science and Numerical Simulation, 44*, 362–380.
Luo, A. C. J., & Thapa, S. (2009). Periodic motions in a simplified brake system with a periodic excitation. *Communications in Nonlinear Science and Numerical Simulation, 128*, 97–105.
Min, F. H., & Luo, A. C. J. (2015). Complex dynamics of projective synchronization of chua circuits with different scrolls. *International Journal of Bifurcation and Chaos, 25*, 1530016.
Min, F. H., & Ma, H. Y. (2020). The mechanism of switching combination synchronization for three distinct nonautonomous systems under sinusoidal constraints. *Nonlinear Dynamics, 100*, 475–492.
Min, F. H., & Zhang, W. (2021). Switching dynamics of a non-autonomous FitzHugh-Nagumo circuit with piecewise-linear flux-controlled memristor. *Chaos, Solitons & Fractals, 152*, 111369.
Poincaré, H. (1892). *Les Methods Nouvelles de la Mecanique Celeste* (Vol. 1). Gauthier-Villars.
Utkin, V. I. (1977). Variable structure systems with sliding modes. *IEEE Transactions on Automatic Control, AC-22*, 212–222.
Utkin, V. I. (1978). *Sliding modes and their application in variable structure systems*. Mir.
Utkin, V. I. (1981). *Sliding regimes in optimization and control problem*. Nauka.
Zhusubaliyev, Z., & Mosekilde, E. (2003). *Bifurcations and chaos in piecewise-smooth dynamical systems*. World Scientific.

Chapter 2
The Flow Passability Theory

Contents

2.1 Basic Concepts 9
2.2 The Flow Passability Theory 13
References 21

In this chapter, a general theory for the passability of a flow to a specific boundary in discontinuous dynamical systems is reviewed from Luo (2009, 2012). The G-functions for discontinuous dynamical systems is presented. From the G-function, the passability of a flow from a domain to an adjacent one is given from Luo (2008, 2009, 2012). The semi-passable flows, non-passable flows (sink flows and source flows), and grazing flows are reviewed.

2.1 Basic Concepts

In 2000–2012, Luo (2005, 2006, 2008, 2009, 2012) developed the flow passability theory in discontinuous dynamical systems. From the Luo's discontinuous dynamical system theory, the following definitions and theorems are stated for reviews.

From Luo (2005, 2006, 2008, 2009, 2012), the domain accessibility definitions are given as follows.

Definition 2.1 A sub-domain in the universal domain $\mho$ in a discontinuous dynamical system is termed an *accessible* sub-domain if at least a specific, continuous vector field can be defined on such a sub-domain.

Definition 2.2 A sub-domain in a universal domain $\mho$ in discontinuous dynamical systems is termed an *inaccessible* sub-domain if no any vector fields can be defined on such a sub-domain.

F. Min, *Discontinuous Dynamics and System Synchronization*, Nonlinear Systems and Complexity 39, https://doi.org/10.1007/978-3-031-66648-3_2

Definition 2.3 On the αth open sub-domain Ω_α, there is a C^{r_α} -continuous system ($r_\alpha \geq 1$) in the form of

$$\dot{\mathbf{x}}^{(\alpha)} \equiv \mathbf{F}^{(\alpha)}\left(\mathbf{x}^{(\alpha)}, t, \mathbf{p}_\alpha\right) \in \mathscr{R}^n, \mathbf{x}^{(\alpha)} = \left(x_1{}^{(\alpha)}, x_2{}^{(\alpha)}, \ldots x_n{}^{(\alpha)}\right)^T \in \Omega_\alpha. \tag{2.1}$$

The time is t and $\dot{\mathbf{x}} - d\mathbf{x}/dt$. In an accessible sub-domain Ω_α, the vector field $\mathbf{F}^{(\alpha)}(\mathbf{x}, t, \mathbf{p}_\alpha)$ with parameter vector $\mathbf{p}_\alpha = \left(p_\alpha^{(1)}, p_\alpha^{(2)}, \ldots p_\alpha^{(l)}\right)^T \in \mathscr{R}^l$ is C^{r_α} continuous ($r_\alpha \geq 1$) in $\mathbf{x} \in \Omega_\alpha$ and for all time t; and the continuous flow in Eq. 2.1 $\mathbf{x}^{(\alpha)}(t) = \mathbf{\Phi}^{(\alpha)}(\mathbf{x}^{(\alpha)}(t), t_0\mathbf{p}_\alpha)$ with $\mathbf{x}^{(\alpha)}(t_0) = \mathbf{\Phi}^{(\alpha)}(\mathbf{x}^{(\alpha)}(t_0), t_0\mathbf{p}_\alpha)$ is C^{r+1} continuous for time t.

Definition 2.4 The boundary in n-dimensional phase space is defined as

$$\begin{aligned} S_{ij} &\equiv \partial\Omega_{ij} = \bar{\Omega}_i \cap \bar{\Omega}_j \\ &= \{\mathrm{x} | \varphi_{ij}(\mathbf{x}, t, \lambda) = 0, \varphi_{ij} \text{ is } \mathrm{C}^r \text{ -- continuous } (r \geq 1)\} \subset \mathscr{R}^{n-1}. \end{aligned} \tag{2.2}$$

Definition 2.5 The two sub-domains, Ω_i and Ω_j, are *disjoint* if the boundary $\partial\Omega_{ij}$ is an empty set (i.e., $\partial\Omega_{ij} = \varnothing$).

Definition 2.6 The flow on the boundary $\partial\Omega_{ij}$ is determined by

$$\dot{\mathbf{x}}^{(0)} = \mathbf{F}^{(0)}(\mathbf{x}^{(0)}, t) \text{ with } \varphi_{ij}(\mathbf{x}^{(0)}, t, \lambda) = 0 \tag{2.3}$$

where $\mathbf{x}^{(0)} = \left(x_1^{(0)}, x_2^{(0)}, \ldots, x_n^{(0)}\right)^{\mathrm{T}}$. With specific initial conditions, one always obtains different flows on $\varphi_{ij}(\mathbf{x}^{(0)}, t, \lambda) = \varphi_{ij}(\mathbf{x}_0^{(0)}, t_0, \lambda) = 0$.

From Luo (2009, 2012), the zero-order G-functions are introduced as follows.

Definition 2.7 Consider a dynamic system in Eq. 2.1 in domain Ω_α ($\alpha \in \{i, j\}$) which has a flow $\mathbf{x}^{(\alpha)} = \mathbf{\Phi}\left(t_0, \mathbf{x}_0^{(\alpha)}, \mathbf{p}_\alpha, t\right)$, and on the boundary $\partial\Omega_{ij}$, there is an enough smooth flow $\mathbf{x}^{(0)} = \mathbf{\Phi}\left(t_0, \mathbf{x}_0^{(0)}, \boldsymbol{\lambda}, t\right)$. For an arbitrarily small $\varepsilon > 0$, there are two time intervals $[t - \varepsilon, t)$ or $(t, t + \varepsilon]$ for flow $\mathbf{x}^{(\alpha)}$ ($\alpha \in \{i, j\}$). The G-functions $G_{\partial\Omega_{ij}}^{(\alpha)}$ of the domain flow $\mathbf{x}^{(\alpha)}$ to the boundary flow $\mathbf{x}^{(0)}$ on the boundary in the normal direction of the boundary $\partial\Omega_{ij}$ are defined as

$$\begin{aligned} & G_{\partial\Omega_{ij}}^{(\alpha)}\left(\mathbf{x}_t^{(0)}, t_-, \mathbf{x}_{t_-}^{(\alpha)}, \mathbf{p}_\alpha, \boldsymbol{\lambda}\right) \\ &= \lim_{\varepsilon \to 0} \frac{1}{\varepsilon}\left[{}^{t}\mathbf{n}_{\partial\Omega_{ij}}^{\mathrm{T}} \cdot \left(\mathbf{x}_{t_-}^{(\alpha)} - \mathbf{x}_t^{(0)}\right) - {}^{t-\varepsilon}\mathbf{n}_{\partial\Omega_{ij}}^{\mathrm{T}} \cdot \left(\mathbf{x}_{t-\varepsilon}^{(\alpha)} - \mathbf{x}_{t-\varepsilon}^{(0)}\right)\right], \\ & G_{\partial\Omega_{ij}}^{(\alpha)}\left(\mathbf{x}_t^{(0)}, t_+, \mathbf{x}_{t_+}^{(\alpha)}, \mathbf{p}_\alpha, \boldsymbol{\lambda}\right) \\ &= \lim_{\varepsilon \to 0} \frac{1}{\varepsilon}\left[{}^{t+\varepsilon}\mathbf{n}_{\partial\Omega_{ij}}^{\mathrm{T}} \cdot \left(\mathbf{x}_{t+\varepsilon}^{(\alpha)} - \mathbf{x}_{t+\varepsilon}^{(0)}\right) - {}^{t}\mathbf{n}_{\partial\Omega_{ij}}^{\mathrm{T}} \cdot \left(\mathbf{x}_{t_+}^{(\alpha)} - \mathbf{x}_t^{(0)}\right)\right]. \end{aligned} \tag{2.4}$$

From Eq. 2.4 with the solutions of Eqs. 2.1 and 2.3, the G-function in Luo (2008, 2009, 2012) is given by

$$
\begin{aligned}
& G_{\partial\Omega_{ij}}^{(\alpha)}\left(\mathbf{x}_t^{(0)}, t_{t\pm}, \mathbf{x}_{t\pm}^{(\alpha)}, \mathbf{p}_\alpha, \boldsymbol{\lambda}\right) \\
& = D_0{}^t\mathbf{n}_{\partial\Omega_{ij}}^{\mathrm{T}} \cdot \left(\mathbf{x}_{t\pm}^{(\alpha)} - \mathbf{x}_t^{(0)}\right) + {}^t\mathbf{n}_{\partial\Omega_{ij}}^{\mathrm{T}} \cdot \left(\dot{\mathbf{x}}_{t\pm}^{(\alpha)} - \dot{\mathbf{x}}_t^{(0)}\right)
\end{aligned}
\tag{2.5}
$$

where the total derivative operators are defined as

$$
D_0(\cdot) \equiv \frac{\partial(\cdot)}{\partial \mathbf{x}^{(0)}}\dot{\mathbf{x}}^{(0)} + \frac{\partial(\cdot)}{\partial t} \text{ and } D_\alpha(\cdot) \equiv \frac{\partial(\cdot)}{\partial \mathbf{x}^{(\alpha)}}\dot{\mathbf{x}}^{(\alpha)} + \frac{\partial(\cdot)}{\partial t}. \tag{2.6}
$$

Using Eqs. 2.1 and 2.4, the *G-function* in Eq. 2.5 becomes

$$
\begin{aligned}
& G_{\partial\Omega_{ij}}^{(\alpha)}\left(\mathbf{x}_t^{(0)}, t_\pm, \mathbf{x}_{t_\pm}^{(\alpha)}, \mathbf{p}_\alpha, \boldsymbol{\lambda}\right) \\
& = D_0{}^t\mathbf{n}_{\partial\Omega_{ij}}^{\mathrm{T}} \cdot \left(\mathbf{x}_{t_\pm}^{(\alpha)} - \mathbf{x}_t^{(0)}\right) + {}^t\mathbf{n}_{\partial\Omega_{ij}}^{\mathrm{T}} \cdot \left[\mathbf{F}^{(\alpha)}\left(\mathbf{x}_{t_\pm}^{(\alpha)}, t_\pm, \mathbf{p}_\alpha\right) - \mathbf{F}^{(0)}\left(\mathbf{x}_t^{(0)}, t, \boldsymbol{\lambda}\right)\right].
\end{aligned}
\tag{2.7}
$$

Consider the flow contacting with the boundary at time t_m (i.e., $\mathbf{x}_m^{(\alpha)} = \mathbf{x}_m^{(0)}$). Because a flow $\mathbf{x}^{(\alpha)}(t)$ approaches the separation boundary with the zero-order $\mathbf{x}^{(\alpha)}(t_{m\pm}) = \mathbf{x}_m = \mathbf{x}^{(0)}(t_m)$, the G-function becomes

$$
\begin{aligned}
& G_{\partial\Omega_{ij}}^{(\alpha)}(\mathbf{x}_m, t_{m\pm}, \mathbf{p}_\alpha, \boldsymbol{\lambda}) \\
& \quad \equiv \mathbf{n}_{\partial\Omega_{ij}}^{\mathrm{T}}\left(\mathbf{x}^{(0)}, t, \boldsymbol{\lambda}\right) \cdot \left[\dot{\mathbf{x}}^{(\alpha)}(t) - \dot{\mathbf{x}}^{(0)}(t)\right]\Big|_{\left(\mathbf{x}_m^{(0)}, \mathbf{x}_{m\pm}^{(\alpha)}, t_{m\pm}\right)} \\
& \quad = \left[\mathbf{n}_{\partial\Omega_{ij}}^{\mathrm{T}}\left(\mathbf{x}^{(0)}, t, \boldsymbol{\lambda}\right) \cdot \dot{\mathbf{x}}^{(\alpha)}(t) + \frac{\partial\varphi_{ij}(\mathbf{x}^{(0)}, t, \boldsymbol{\lambda})}{\partial t}\right]\Bigg|_{\left(\mathbf{x}_m^{(0)}, \mathbf{x}_{m\pm}^{(\alpha)}, t_{m\pm}\right)} \\
& \quad = \left[\nabla\varphi_{ij}\left(\mathbf{x}^{(0)}, t, \boldsymbol{\lambda}\right) \cdot \dot{\mathbf{x}}^{(\alpha)}(t) + \frac{\partial\varphi_{ij}(\mathbf{x}^{(0)}, t, \boldsymbol{\lambda})}{\partial t}\right]\Bigg|_{\left(\mathbf{x}_m^{(0)}, \mathbf{x}_{m\pm}^{(\alpha)}, t_{m\pm}\right)}.
\end{aligned}
\tag{2.8}
$$

With Eqs. 2.1 and 2.5, Equation 2.8 becomes

$$
\begin{aligned}
& G_{\partial\Omega_{ij}}^{(\alpha)}(\mathbf{x}_m, t_{m\pm}, \mathbf{p}_\alpha, \boldsymbol{\lambda}) \\
& \quad = \mathbf{n}_{\partial\Omega_{ij}}^{\mathrm{T}}\left(\mathbf{x}^{(0)}, t, \boldsymbol{\lambda}\right) \cdot \left[\mathbf{F}(\mathbf{x}^{(\alpha)}, t, \mathbf{p}_\alpha) - \mathbf{F}^{(0)}\left(\mathbf{x}^{(0)}, t, \boldsymbol{\lambda}\right)\right]\Big|_{\left(\mathbf{x}_m^{(0)}, \mathbf{x}_{m\pm}^{(\alpha)}, t_m \pm\right)} \\
& \quad = \left[\mathbf{n}_{\partial\Omega_{ij}}^{\mathrm{T}}\left(\mathbf{x}^{(0)}, t, \boldsymbol{\lambda}\right) \cdot \mathbf{F}(\mathbf{x}^{(\alpha)}, t, \mathbf{p}_\alpha) + \frac{\partial\varphi_{ij}(\mathbf{x}^{(0)}, t, \boldsymbol{\lambda})}{\partial t}\right]\Bigg|_{\left(\mathbf{x}_m^{(0)}, \mathbf{x}_{m\pm}^{(\alpha)}, t_m \pm\right)} \\
& \quad = \left[\nabla\varphi_{ij}\left(\mathbf{x}^{(0)}, t, \boldsymbol{\lambda}\right) \cdot \mathbf{F}(\mathbf{x}^{(\alpha)}, t, \mathbf{p}_\alpha) + \frac{\partial\varphi_{ij}(\mathbf{x}^{(0)}, t, \boldsymbol{\lambda})}{\partial t}\right]\Bigg|_{\left(\mathbf{x}_m^{(0)}, \mathbf{x}_{m\pm}^{(\alpha)}, t_m \pm\right)}.
\end{aligned}
\tag{2.9}
$$

From Luo (2009, 2012), the higher-order G-functions are introduced as follows.

Definition 2.8 Consider a dynamic system in Eq. 2.1 in domain Ω_α ($\alpha \in \{i,j\}$) which has the flow $\mathbf{x}_t^{(\alpha)} = \boldsymbol{\Phi}\left(t_0, \mathbf{x}_0^{(\alpha)}, \mathbf{p}_\alpha, t\right)$, and on the boundary $\partial\Omega_{ij}$, there is an enough smooth flow $\mathbf{x}_t^{(0)} = \boldsymbol{\Phi}\left(t_0, \mathbf{x}_0^{(0)}, \boldsymbol{\lambda}, t\right)$. For an arbitrarily small $\varepsilon > 0$, there are two time intervals $[t-\varepsilon, t)$ and $(t, t+\varepsilon]$ for a domain flow $\mathbf{x}_t^{(\alpha)}$ ($\alpha \in \{i, j\}$). The vector fields $\mathbf{F}^{(\alpha)}(\mathbf{x}^{(\alpha)}, t, \mathbf{p}_\alpha)$ and $\mathbf{F}^{(0)}(\mathbf{x}^{(0)}, t, \boldsymbol{\lambda})$ are $C^{r_\alpha}_{[t-\varepsilon,t+\varepsilon]}$-continuous ($r_\alpha \geq k$) for time t with $\left\| d^{r_\alpha+1}\mathbf{x}_t^{(\alpha)}/dt^{r_\alpha+1} \right\| < \infty$ and $\left\| d^{r_\alpha+1}\mathbf{x}_t^{(0)}/dt^{r_\alpha+1} \right\| < \infty$. The kth-*order G-functions* of the domain flow $\mathbf{x}_t^{(\alpha)}$ to the boundary flow $\mathbf{x}_t^{(0)}$ in the normal direction of $\partial\Omega_{ij}$ are defined as

$$
\begin{aligned}
&G^{(k,\alpha)}_{\partial\Omega_{ij}}\left(\mathbf{x}_t^{(0)}, t_-, \mathbf{x}_{t_-}^{(\alpha)}, \mathbf{p}_\alpha, \boldsymbol{\lambda}\right) \\
&= \lim_{\varepsilon \to 0} \frac{(-1)^{k+2}}{\varepsilon^{k+1}} \left[{}^t\mathbf{n}^{\mathrm{T}}_{\partial\Omega_{ij}} \cdot \left(\mathbf{x}_{t_-}^{(\alpha)} - \mathbf{x}_t^{(0)}\right) - {}^{t-\varepsilon}\mathbf{n}^{\mathrm{T}}_{\partial\Omega_{ij}} \cdot \left(\mathbf{x}_{t-\varepsilon}^{(\alpha)} - \mathbf{x}_{t-\varepsilon}^{(0)}\right) \right. \\
&\quad \left. + \sum_{s=0}^{k-1} G^{(s,\alpha)}_{\partial\Omega_{ij}}\left(\mathbf{x}_t^{(0)}, t, \mathbf{x}_{t_-}^{(\alpha)}, \mathbf{p}_\alpha, \boldsymbol{\lambda}\right)(-\varepsilon)^{s+1} \right] \\
&= \lim_{\varepsilon \to 0} \frac{1}{\varepsilon^{k+1}} \left[{}^{t+\varepsilon}\mathbf{n}^{\mathrm{T}}_{\partial\Omega_{ij}} \cdot \left(\mathbf{x}_{t+\varepsilon}^{(\alpha)} - \mathbf{x}_{t+\varepsilon}^{(0)}\right) - {}^t\mathbf{n}^{\mathrm{T}}_{\partial\Omega_{ij}} \cdot \left(\mathbf{x}_{t_+}^{(\alpha)} - \mathbf{x}_t^{(0)}\right) \right. \\
&\quad \left. - \sum_{s=0}^{k-1} G^{(s,\alpha)}_{\partial\Omega_{ij}}\left(\mathbf{x}_t^{(0)}, t, \mathbf{x}_{t_+}^{(\alpha)}, \mathbf{p}_\alpha, \boldsymbol{\lambda}\right)\varepsilon^{s+1} \right].
\end{aligned}
\tag{2.10}
$$

Then, the Taylor series expansion applying to Eq. 2.10 with Eqs. 2.1 and 2.3, the kth-order G-function of the flow $\mathbf{x}_t^{(\alpha)}$ to the boundary $\partial\Omega_{ij}$ as in Luo (2008, 2009, 2012) is

$$
\begin{aligned}
&G^{(k,\alpha)}_{\partial\Omega_{ij}}\left(\mathbf{x}_t^{(0)}, t_\pm, \mathbf{x}_{t_\pm}^{(\alpha)}, \mathbf{p}_\alpha, \boldsymbol{\lambda}\right) \\
&= \sum_{s=1}^{k+1} C^s_{k+1} D_0^{k+1-s}\, {}^t\mathbf{n}^{\mathrm{T}}_{\partial\Omega_{ij}} \cdot \left[D_\alpha^{s-1}\mathbf{F}^{(\alpha)}\left(\mathbf{x}^{(\alpha)}, t, \mathbf{p}_\alpha\right) \right. \\
&\quad \left. - D_0^{s-1}\mathbf{F}^{(0)}\left(\mathbf{x}^{(0)}, t, \boldsymbol{\lambda}\right) \right] \Big|_{\left(\mathbf{x}_t^{(0)}, \mathbf{x}_{t_\pm}^{(\alpha)}, t_\pm\right)} + D_0^{k+1}\, {}^t\mathbf{n}^{\mathrm{T}}_{\partial\Omega_{ij}} \cdot \left(\mathbf{x}_{t_\pm}^{(\alpha)} - \mathbf{x}_t^{(0)}\right),
\end{aligned}
\tag{2.11}
$$

where

$$
C^s_{k+1} = \frac{(k+1)k(k-1)\cdots(k+2-s)}{s!} \tag{2.12}
$$

with $C^0_{k+1} = 1$ and $s\,! = 1 \times 2 \times \cdots \times s$.

If a flow contacting with $\partial\Omega_{ij}$ at time t_m (i.e., $\mathbf{x}_{m\pm}^{(\alpha)} = \mathbf{x}_m^{(0)}$) and ${}^t\mathbf{n}^{\mathrm{T}}_{\partial\Omega_{ij}} \equiv \mathbf{n}^{\mathrm{T}}_{\partial\Omega_{ij}}$, the kth-order G-function in Luo (2008, 2009, 2012) is

$$
\begin{aligned}
& G_{\partial\Omega_{ij}}^{(k,\alpha)}(\mathbf{x}_m, t_{m\pm}, \mathbf{p}_\alpha, \boldsymbol{\lambda}) \\
&= \sum\nolimits_{r=1}^{k+1} C_{k+1}^r D_0^{k+1-r} \mathbf{n}_{\partial\Omega_{ij}}^{\mathrm{T}} \cdot \left[\frac{d^r \mathbf{x}^{(\alpha)}}{dt^r} - \frac{d^r \mathbf{x}^{(0)}}{dt^r} \right] \Bigg|_{(\mathbf{x}_m^{(0)}, \mathbf{x}_{m\pm}^{(\alpha)}, t_{m\pm})} \\
&= \sum\nolimits_{r=1}^{k+1} C_{k+1}^r D_0^{k+1-r} \mathbf{n}_{\partial\Omega_{ij}}^{\mathrm{T}} \cdot \left[D_\alpha^{r-1} \mathbf{F}\left(\mathbf{x}^{(\alpha)}, t, \mathbf{p}_\alpha\right) - D_0^{r-1} \mathbf{F}^{(0)}\left(\mathbf{x}^{(0)} t, \boldsymbol{\lambda}\right) \right] \Bigg|_{(\mathbf{x}_m^{(0)}, \mathbf{x}_{m\pm}^{(\alpha)}, t_{m\pm})}.
\end{aligned}
\tag{2.13}
$$

2.2 The Flow Passability Theory

From the flow passability theory in Luo (2008, 2009, 2012), the flow passability on the boundary in discontinuous dynamical systems is defined as follows.

Definition 2.9 For a discontinuous dynamical system in Eq. 2.1, there is a point $\mathbf{x}^{(0)}(t_m) \equiv \mathbf{x}_m \in \partial\Omega_{ij}$ at time t_m between two adjacent domains Ω_α $(\alpha = i, j)$. For an arbitrarily small $\varepsilon > 0$, there are two time intervals $[t_{m-\varepsilon}, t_m)$ and $(t_m, t_{m+\varepsilon}]$. Suppose $\mathbf{x}^{(i)}(t_{m-}) = \mathbf{x}_m = \mathbf{x}^{(j)}(t_{m+})$. The flow $\mathbf{x}^{(i)}(t)$ and $\mathbf{x}^{(j)}(t)$ to the boundary $\partial\Omega_{ij}$ is *semi-passable* from domain Ω_i to Ω_j if

$$
\begin{aligned}
&\text{either} \left.\begin{aligned}
&\mathbf{n}_{\partial\Omega_{ij}}^{\mathrm{T}}\left(\mathbf{x}_{m-\varepsilon}^{(0)}\right) \cdot \left[\mathbf{x}_{m-\varepsilon}^{(0)} - \mathbf{x}_{m-\varepsilon}^{(i)}\right] > 0 \\
&\mathbf{n}_{\partial\Omega_{ij}}^{\mathrm{T}}\left(\mathbf{x}_{m+\varepsilon}^{(0)}\right) \cdot \left[\mathbf{x}_{m+\varepsilon}^{(j)} - \mathbf{x}_{m+\varepsilon}^{(0)}\right] > 0
\end{aligned}\right\} \text{for } \mathbf{n}_{\partial\Omega_{ij}} \to \Omega_j, \\
&\text{or} \left.\begin{aligned}
&\mathbf{n}_{\partial\Omega_{ij}}^{\mathrm{T}}\left(\mathbf{x}_{m-\varepsilon}^{(0)}\right) \cdot \left[\mathbf{x}_{m-\varepsilon}^{(0)} - \mathbf{x}_{m-\varepsilon}^{(i)}\right] < 0 \\
&\mathbf{n}_{\partial\Omega_{ij}}^{\mathrm{T}}\left(\mathbf{x}_{m+\varepsilon}^{(0)}\right) \cdot \left[\mathbf{x}_{m+\varepsilon}^{(j)} - \mathbf{x}_{m+\varepsilon}^{(0)}\right] < 0
\end{aligned}\right\} \text{for } \mathbf{n}_{\partial\Omega_{ij}} \to \Omega_i.
\end{aligned}
\tag{2.14}
$$

The analytical conditions for such a passable flow on $\partial\Omega_{ij}$ from domain Ω_i to Ω_j are given in Luo (2008, 2009, 2012), which can be stated in the following theorem.

Theorem 2.1 *For a discontinuous dynamical system in Eq.* 2.1, *there is a point* $\mathbf{x}^{(0)}(t_m) \equiv \mathbf{x}_m \in \partial\Omega_{ij}$ *at time* t_m *between two adjacent domains* Ω_α $(\alpha = i, j)$. *For an arbitrarily small* $\varepsilon > 0$, *there are two time intervals* $[t_{m-\varepsilon}, t_m)$ *and* $(t_m, t_{m+\varepsilon}]$. *Suppose* $\mathbf{x}^{(i)}(t_{m-}) = \mathbf{x}_m = \mathbf{x}^{(j)}(t_{m+})$. *Two flows* $\mathbf{x}^{(i)}(t)$ *and* $\mathbf{x}^{(j)}(t)$ *are* $C_{[t_{m-\varepsilon}, t_m)}^{r_i}$ *and* $C_{(t_m, t_{m+\varepsilon}]}^{r_j}$*-continuous for time t, respectively, and* $\left\| d^{r_\alpha+1}\mathbf{x}^{(\alpha)}/dt^{r_\alpha+1} \right\| < \infty$ $(r_\alpha \geq 1$, $\alpha = i, j)$. *The flow* $\mathbf{x}^{(i)}(t)$ *and* $\mathbf{x}^{(j)}(t)$ *to the boundary* $\partial\Omega_{ij}$ *is semi-passable from domain* Ω_i *to* Ω_j *if and only if*

$$\left.\begin{array}{ll} \textit{either} & \left.\begin{array}{l} G^{(i)}_{\partial\Omega_{ij}}(\mathbf{x}_m, t_{m-}, \mathbf{p}_i, \boldsymbol{\lambda}) > 0 \\ G^{(j)}_{\partial\Omega_{ij}}(\mathbf{x}_m, t_{m+}, \mathbf{p}_j, \boldsymbol{\lambda}) > 0 \end{array}\right\} \textit{for } \mathbf{n}_{\partial\Omega_{ij}} \rightarrow \Omega_j, \\ \textit{or} & \left.\begin{array}{l} G^{(i)}_{\partial\Omega_{ij}}(\mathbf{x}_m, t_{m-}, \mathbf{p}_i, \boldsymbol{\lambda}) < 0 \\ G^{(j)}_{\partial\Omega_{ij}}(\mathbf{x}_m, t_{m+}, \mathbf{p}_j, \boldsymbol{\lambda}) < 0 \end{array}\right\} \textit{for } \mathbf{n}_{\partial\Omega_{ij}} \rightarrow \Omega_i. \end{array}\right. \tag{2.15}$$

Proof See the proof in Luo (Luo, 2009; Luo, 2012; Luo, 2005).

Definition 2.10 For a discontinuous dynamical system in Eq. 2.1, there is a point $\mathbf{x}^{(0)}(t_m) \equiv \mathbf{x}_m \in \partial\Omega_{ij}$ at time t_m between two adjacent domains Ω_α ($\alpha = i, j$). For an arbitrarily small $\varepsilon > 0$, there are two time intervals $[t_{m-\varepsilon}, t_m)$ and $(t_m, t_{m+\varepsilon}]$. Suppose $\mathbf{x}^{(i)}(t_{m-}) = \mathbf{x}_m = \mathbf{x}^{(j)}(t_{m+})$. A flow $\mathbf{x}^{(i)}(t)$ is $C^{r_i}_{[t_{m-\varepsilon}, t_m)}$ continuous for time t with $\left\| d^{r_i+1}\mathbf{x}^{(i)}/dt^{r_i+1} \right\| < \infty$ ($r_i \geq 2k_i + 1$), and a flow $\mathbf{x}^{(j)}(t)$ is $C^{r_j}_{(t_m, t_{m+\varepsilon}]}$-continuous with $\left\| d^{r_j+1}\mathbf{x}^{(j)}/dt^{r_j+1} \right\| < \infty$ ($r_j \geq m_j + 1$). The flow $\mathbf{x}^{(i)}(t)$ of the $(2k_i)$th-order and $\mathbf{x}^{(j)}(t)$ of the (m_j)th-order to the boundary $\partial\Omega_{ij}$ is $(2k_i : m_j)$-*semi-passable* from domain Ω_i to Ω_j if

$$\begin{array}{l} G^{(s,i)}_{\partial\Omega_{ij}}(\mathbf{x}_m, t_{m-}, \mathbf{p}_i, \boldsymbol{\lambda}) = 0 \ \text{ for } s = 0, 1, \cdots, 2k_i - 1, \\ G^{(2k_i,i)}_{\partial\Omega_{ij}}(\mathbf{x}_m, t_{m-}, \mathbf{p}_i, \boldsymbol{\lambda}) \neq 0, \end{array} \tag{2.16}$$

$$\begin{array}{l} G^{(s,j)}_{\partial\Omega_{ij}}(\mathbf{x}_m, t_{m+}, \mathbf{p}_j, \boldsymbol{\lambda}) = 0 \text{ for } s = 0, 1, \cdots, m_j - 1, \\ G^{(m_j,j)}_{\partial\Omega_{ij}}(\mathbf{x}_m, t_{m+}, \mathbf{p}_j, \boldsymbol{\lambda}) \neq 0, \end{array} \tag{2.17}$$

$$\left.\begin{array}{ll} \text{either} & \left.\begin{array}{l} \mathbf{n}^{\mathrm{T}}_{\partial\Omega_{ij}}(\mathbf{x}^{(0)}_{m-\varepsilon}) \cdot [\mathbf{x}^{(0)}_{m-\varepsilon} - \mathbf{x}^{(i)}_{m-\varepsilon}] > 0 \\ \mathbf{n}^{\mathrm{T}}_{\partial\Omega_{ij}}\left(\mathbf{x}^{(0)}_{m+\varepsilon}\right) \cdot \left[\mathbf{x}^{(j)}_{m+\varepsilon} - \mathbf{x}^{(0)}_{m+\varepsilon}\right] > 0 \end{array}\right\} \text{for } \mathbf{n}_{\partial\Omega_{ij}} \rightarrow \Omega_j, \\ \text{or} & \left.\begin{array}{l} \mathbf{n}^{\mathrm{T}}_{\partial\Omega_{ij}}(\mathbf{x}^{(0)}_{m-\varepsilon}) \cdot [\mathbf{x}^{(0)}_{m-\varepsilon} - \mathbf{x}^{(i)}_{m-\varepsilon}] < 0 \\ \mathbf{n}^{\mathrm{T}}_{\partial\Omega_{ij}}\left(\mathbf{x}^{(0)}_{m+\varepsilon}\right) \cdot \left[\mathbf{x}^{(j)}_{m+\varepsilon} - \mathbf{x}^{(0)}_{m+\varepsilon}\right] < 0 \end{array}\right\} \text{for } \mathbf{n}_{\partial\Omega_{ij}} \rightarrow \Omega_i. \end{array}\right. \tag{2.18}$$

The analytical conditions for such a $(2k_i : m_j)$ *semi*-passable flow on $\partial\Omega_{ij}$ from domain Ω_i to Ω_j are given in Luo (2005, 2009, 2012), which can be stated in the following theorem.

Theorem 2.2 *For a discontinuous dynamical system in Eq. 2.1, there is a point* $\mathbf{x}^{(0)}(t_m) \equiv \mathbf{x}_m \in \partial\Omega_{ij}$ *at time* t_m *between two adjacent domains* Ω_α ($\alpha = i, j$). *For an arbitrarily small* $\varepsilon > 0$, *there are two time intervals* $[t_{m-\varepsilon}, t_m)$ *and* $(t_m, t_{m+\varepsilon}]$. *Suppose* $\mathbf{x}^{(i)}(t_{m-}) = \mathbf{x}_m = \mathbf{x}^{(j)}(t_{m+})$. *A flow* $\mathbf{x}^{(i)}(t)$ *is* $C^{r}_{[t_{m-\varepsilon}, t_m)}$ *continuous for time t with* $\left\| d^{r_i+1}\mathbf{x}^{(i)}/dt^{r_i+1} \right\| < \infty$ ($r_i \geq 2k_i + 1$), *and a flow* $\mathbf{x}^{(j)}(t)$ *is* $C^{r_j}_{(t_m, t_{m+\varepsilon}]}$*-continuous* *with* $\left\| d^{r_j+1}\mathbf{x}^{(j)}/dt^{r_j+1} \right\| < \infty$ $(r_j \geq m_j + 1)$. *The flow* $\mathbf{x}^{(i)}(t)$ *of the* $(2k_i)$th-*order and* $\mathbf{x}^{(j)}(t)$ *of the* (m_j)th-*order to the boundary* $\partial\Omega_{ij}$ *is* $(2k_i : m_j)$-*semi-passable from domain* Ω_i *to* Ω_j *if and only if*

$$G^{(s,i)}_{\partial\Omega_{ij}}(\mathbf{x}_m, t_{m-}, \mathbf{p}_i, \boldsymbol{\lambda}) = 0 \ \textit{for}\ s = 0, 1, \cdots, 2k_i - 1; \tag{2.19}$$

$$G^{(s,j)}_{\partial\Omega_{ij}}(\mathbf{x}_m, t_{m+}, \mathbf{p}_j, \boldsymbol{\lambda}) = 0 \ \textit{for}\ s = 0, 1, \cdots, m_j - 1; \tag{2.20}$$

$$\begin{array}{ll} \textit{either} & \left.\begin{array}{l} G^{(2k_i,i)}_{\partial\Omega_{ij}}(\mathbf{x}_m, t_{m-}, \mathbf{p}_i, \boldsymbol{\lambda}) > 0 \\ G^{(m_j,j)}_{\partial\Omega_{ij}}(\mathbf{x}_m, t_{m+}, \mathbf{p}_j, \boldsymbol{\lambda}) > 0 \end{array}\right\} \textit{for}\ \mathbf{n}_{\partial\Omega_{ij}} \to \Omega_j, \\ \textit{or} & \left.\begin{array}{l} G^{(2k_i,i)}_{\partial\Omega_{ij}}(\mathbf{x}_m, t_{m-}, \mathbf{p}_i, \boldsymbol{\lambda}) < 0 \\ G^{(m_j,j)}_{\partial\Omega_{ij}}(\mathbf{x}_m, t_{m+}, \mathbf{p}_j, \boldsymbol{\lambda}) < 0 \end{array}\right\} \textit{for}\ \mathbf{n}_{\partial\Omega_{ij}} \to \Omega_i. \end{array} \tag{2.21}$$

Proof See the proof in Luo (2008, 2009, 2012).

As in Luo (2008, 2009, 2012), the non-passable flows are called the *full non-passable flows* because the flows on both sides of the boundary will approach or leave the boundary. The non-passable flow of the first kind is defined as follows.

Definition 2.11 For a discontinuous dynamic system in Eq. 2.1, there is a point $\mathbf{x}^{(0)}(t_m) \equiv \mathbf{x}_m \in \partial\Omega_{ij}$ at time t_m between two adjacent domains Ω_α $(\alpha = i, j)$. For an arbitrarily small $\varepsilon > 0$, there is a time interval $[t_{m-\varepsilon}, t_m)$. Suppose $\mathbf{x}^{(\alpha)}(t_{m-}) = \mathbf{x}_m$. The flow $\mathbf{x}^{(i)}(t)$ and $\mathbf{x}^{(j)}(t)$ to the boundary $\partial\Omega_{ij}$ is *non-passable* of the first kind (or called a sink flow) if

$$\begin{array}{ll} \text{either} & \left.\begin{array}{l} \mathbf{n}^{\mathrm{T}}_{\partial\Omega_{ij}}(\mathbf{x}^{(0)}_{m-\varepsilon}) \cdot [\mathbf{x}^{(0)}_{m-\varepsilon} - \mathbf{x}^{(i)}_{m-\varepsilon}] > 0 \\ \mathbf{n}^{\mathrm{T}}_{\partial\Omega_{ij}}(\mathbf{x}^{(0)}_{m-\varepsilon}) \cdot [\mathbf{x}^{(0)}_{m-\varepsilon} - \mathbf{x}^{(j)}_{m-\varepsilon}] < 0 \end{array}\right\} \text{for}\ \mathbf{n}_{\partial\Omega_{ij}} \to \Omega_j, \\ \text{or} & \left.\begin{array}{l} \mathbf{n}^{\mathrm{T}}_{\partial\Omega_{ij}}(\mathbf{x}^{(0)}_{m-\varepsilon}) \cdot [\mathbf{x}^{(0)}_{m-\varepsilon} - \mathbf{x}^{(i)}_{m-\varepsilon}] < 0 \\ \mathbf{n}^{\mathrm{T}}_{\partial\Omega_{ij}}(\mathbf{x}^{(0)}_{m-\varepsilon}) \cdot [\mathbf{x}^{(0)}_{m-\varepsilon} - \mathbf{x}^{(j)}_{m-\varepsilon}] > 0 \end{array}\right\} \text{for}\ \mathbf{n}_{\partial\Omega_{ij}} \to \Omega_i. \end{array} \tag{2.22}$$

The analytical conditions for such a non-passable flow of the first kind on $\partial\Omega_{ij}$ from domain Ω_i to Ω_j are given in Luo (2008, 2009, 2012), which can be stated in the following theorem.

Theorem 2.3 *For a discontinuous dynamical system in Eq. 2.1, there is a point* $\mathbf{x}^{(0)}(t_m) \equiv \mathbf{x}_m \in \partial\Omega_{ij}$ *at time* t_m *between two adjacent domains* Ω_α $(\alpha = i, j)$. *For an arbitrarily small* $\varepsilon > 0$, *there is a time interval* $[t_{m-\varepsilon}, t_m)$. *Suppose* $\mathbf{x}^{(\alpha)}(t_{m-}) = \mathbf{x}_m$. *A flow* $\mathbf{x}^{(\alpha)}(t)$ *is* $C^{r_\alpha}_{[t_{m-\varepsilon}, t_m)}$ *-continuous* $(r_\alpha \geq 1,\ \alpha = i, j)$ *for time* t *with* $\left\| d^{r_\alpha+1}\mathbf{x}^{(\alpha)}/dt^{r_\alpha+1} \right\| < \infty$. *The flow* $\mathbf{x}^{(i)}(t)$ *and* $\mathbf{x}^{(j)}(t)$ *to the boundary* $\partial\Omega_{ij}$ *is non-passable of the first kind (or a sink flow) if and only if*

$$\left.\begin{array}{ll}\textit{either} & \left.\begin{array}{l}G^{(i)}_{\partial\Omega_{ij}}(\mathbf{x}_m, t_{m-}, \mathbf{p}_i, \boldsymbol{\lambda})>0\\ G^{(j)}_{\partial\Omega_{ij}}(\mathbf{x}_m, t_{m-}, \mathbf{p}_j, \boldsymbol{\lambda})<0\end{array}\right\} \textit{for } \mathbf{n}_{\partial\Omega_{ij}}\to\Omega_j,\\ \textit{or} & \left.\begin{array}{l}G^{(i)}_{\partial\Omega_{ij}}(\mathbf{x}_m, t_{m-}, \mathbf{p}_i, \boldsymbol{\lambda})<0\\ G^{(j)}_{\partial\Omega_{ij}}(\mathbf{x}_m, t_{m-}, \mathbf{p}_j, \boldsymbol{\lambda})>0\end{array}\right\} \textit{for } \mathbf{n}_{\partial\Omega_{ij}}\to\Omega_i.\end{array}\right. \tag{2.23}$$

Proof See the proof as in Luo (2008, 2009, 2012). ■

As in Luo (2008, 2009, 2012), the $(2k_i : 2k_j)$-non-passable flow of the first kind is defined as follows.

Definition 2.12 For a discontinuous dynamic system in Eq. 2.1, there is a point $\mathbf{x}^{(0)}(t_m) \equiv \mathbf{x}_m \in \partial\Omega_{ij}$ at time t_m between two adjacent domains Ω_α $(\alpha = i, j)$. For an arbitrarily small $\varepsilon > 0$, there is a time interval $[t_{m-\varepsilon}, t_m)$. Suppose $\mathbf{x}^{(\alpha)}(t_{m-}) = \mathbf{x}_m$. A flow $\mathbf{x}^{(\alpha)}(t)$ is $C^{r_\alpha}_{[t_{m-\varepsilon}, t_m)}$-continuous $(r_\alpha \geq 2k_\alpha + 1,\ \alpha = i, j)$ for time t with $\left\| d^{r_\alpha+1}\mathbf{x}^{(\alpha)}/dt^{r_\alpha+1} \right\| < \infty$. The flow $\mathbf{x}^{(i)}(t)$ of the $(2k_i)$th-order and $\mathbf{x}^{(j)}(t)$ of the $(2k_j)$th-order to the boundary $\partial\Omega_{ij}$ is $(2k_i : 2k_j)$-*non-passable* of the first kind (or called a $(2k_i : 2k_j)$-sink flow) if

$$\left.\begin{array}{l}G^{(s_\alpha,\alpha)}_{\partial\Omega_{ij}}(\mathbf{x}_m, t_{m-}, \mathbf{p}_\alpha, \boldsymbol{\lambda}) = 0 \text{ for } s_\alpha = 0, 1, \cdots, 2k_\alpha - 1,\\ G^{(2k_\alpha,\alpha)}_{\partial\Omega_{ij}}(\mathbf{x}_m, t_{m-}, \mathbf{p}_\alpha, \boldsymbol{\lambda}) \neq 0 \ (\alpha = i, j);\end{array}\right\} \tag{2.24}$$

$$\left.\begin{array}{ll}\text{either} & \left.\begin{array}{l}\mathbf{n}^{\mathrm{T}}_{\partial\Omega_{ij}}(\mathbf{x}^{(0)}_{m-\varepsilon})\cdot[\mathbf{x}^{(0)}_{m-\varepsilon} - \mathbf{x}^{(i)}_{m-\varepsilon}] > 0\\ \mathbf{n}^{\mathrm{T}}_{\partial\Omega_{ij}}(\mathbf{x}^{(0)}_{m-\varepsilon})\cdot[\mathbf{x}^{(0)}_{m-\varepsilon} - \mathbf{x}^{(j)}_{m-\varepsilon}] < 0\end{array}\right\} \text{for } \mathbf{n}_{\partial\Omega_{ij}}\to\Omega_j,\\ \text{or} & \left.\begin{array}{l}\mathbf{n}^{\mathrm{T}}_{\partial\Omega_{ij}}(\mathbf{x}^{(0)}_{m-\varepsilon})\cdot[\mathbf{x}^{(0)}_{m-\varepsilon} - \mathbf{x}^{(i)}_{m-\varepsilon}] < 0\\ \mathbf{n}^{\mathrm{T}}_{\partial\Omega_{ij}}(\mathbf{x}^{(0)}_{m-\varepsilon})\cdot[\mathbf{x}^{(0)}_{m-\varepsilon} - \mathbf{x}^{(j)}_{m-\varepsilon}] > 0\end{array}\right\} \text{for } \mathbf{n}_{\partial\Omega_{ij}}\to\Omega_i.\end{array}\right. \tag{2.25}$$

The analytical conditions for such a $(2k_i : 2k_j)$ non-passable flow of the first kind on $\partial\Omega_{ij}$ from domain Ω_i to Ω_j are given in Luo (2008, 2009, 2012), which can be stated in the following theorem.

Theorem 2.4 *For a discontinuous dynamical system in Eq. 2.1, there is a point* $\mathbf{x}^{(0)}(t_m) \equiv \mathbf{x}_m \in \partial\Omega_{ij}$ *at time* t_m *between two adjacent domains* Ω_α $(\alpha = i, j)$. *For an arbitrarily small* $\varepsilon > 0$, *there is a time interval* $[t_{m-\varepsilon}, t_m)$. *Suppose* $\mathbf{x}^{(\alpha)}(t_{m-}) = \mathbf{x}_m$. *A flow* $\mathbf{x}^{(\alpha)}(t)$ *is* $C^{r_\alpha}_{[t_{m-\varepsilon}, t_m)}$*-continuous* $(r_\alpha \geq 2k_\alpha + 1,\ \alpha = i, j)$ *for time t with* $\left\| d^{r_\alpha+1}\mathbf{x}^{(\alpha)}/dt^{r_\alpha+1} \right\| < \infty$. *The flow* $\mathbf{x}^{(i)}(t)$ *of the* $(2k_i)$th-*order and* $\mathbf{x}^{(j)}(t)$ *of the* $(2k_j)$ th-*order to the boundary* $\partial\Omega_{ij}$ *are* $(2k_i : 2k_j)$*-non-passable of the first kind (or a* $(2k_i : 2k_j)$*-sink flow) if and only if*

$$G^{(s_\alpha,\alpha)}_{\partial\Omega_{ij}}(\mathbf{x}_m, t_{m-}, \mathbf{p}_\alpha, \boldsymbol{\lambda}) = 0 \text{ for } s_\alpha = 0, 1, \cdots, 2k_\alpha - 1 \textit{ and } \alpha = i, j; \tag{2.26}$$

$$\begin{aligned} &\textit{either} \left.\begin{aligned} G^{(2k_i,i)}_{\partial\Omega_{ij}}(\mathbf{x}_m, t_{m-}, \mathbf{p}_i, \boldsymbol{\lambda}) > 0 \\ G^{(2k_j,j)}_{\partial\Omega_{ij}}(\mathbf{x}_m, t_{m-}, \mathbf{p}_j, \boldsymbol{\lambda}) < 0 \end{aligned}\right\} \textit{for } \mathbf{n}_{\partial\Omega_{ij}} \to \Omega_j, \\ &\textit{or} \left.\begin{aligned} G^{(2k_i,i)}_{\partial\Omega_{ij}}(\mathbf{x}_m, t_{m-}, \mathbf{p}_i, \boldsymbol{\lambda}) < 0 \\ G^{(2k_j,j)}_{\partial\Omega_{ij}}(\mathbf{x}_m, t_{m-}, \mathbf{p}_j, \boldsymbol{\lambda}) > 0 \end{aligned}\right\} \textit{for } \mathbf{n}_{\partial\Omega_{ij}} \to \Omega_i. \end{aligned} \tag{2.27}$$

Proof See the proof from Luo (2008, 2009, 2012). ■

From Luo (Luo, 2009; Luo, 2012; Luo, 2008), the definition of the non-passable flow of the second kind is stated as follows.

Definition 2.13 For a discontinuous dynamic system in Eq. 2.1, there is a point $\mathbf{x}^{(0)}(t_m) \equiv \mathbf{x}_m \in \partial\Omega_{ij}$ at time t_m between two adjacent domains Ω_α $(\alpha = i, j)$. For an arbitrarily small $\varepsilon > 0$, there is a time interval $(t_m, t_{m+\varepsilon}]$. Suppose $\mathbf{x}^{(\alpha)}(t_{m+}) = \mathbf{x}_m$. The flow $\mathbf{x}^{(i)}(t)$ and $\mathbf{x}^{(j)}(t)$ to the boundary $\partial\Omega_{ij}$ are *non-passable* of the second kind (or called a source flow) if

$$\begin{aligned} &\text{either} \left.\begin{aligned} \mathbf{n}^{\mathrm{T}}_{\partial\Omega_{ij}}\left(\mathbf{x}^{(0)}_{m+\varepsilon}\right) \cdot \left[\mathbf{x}^{(i)}_{m+\varepsilon} - \mathbf{x}^{(0)}_{m+\varepsilon}\right] < 0 \\ \mathbf{n}^{\mathrm{T}}_{\partial\Omega_{ij}}\left(\mathbf{x}^{(0)}_{m+\varepsilon}\right) \cdot \left[\mathbf{x}^{(j)}_{m+\varepsilon} - \mathbf{x}^{(0)}_{m+\varepsilon}\right] > 0 \end{aligned}\right\} \text{for } \mathbf{n}_{\partial\Omega_{ij}} \to \Omega_j, \\ &\text{or} \left.\begin{aligned} \mathbf{n}^{\mathrm{T}}_{\partial\Omega_{ij}}\left(\mathbf{x}^{(0)}_{m+\varepsilon}\right) \cdot \left[\mathbf{x}^{(i)}_{m+\varepsilon} - \mathbf{x}^{(0)}_{m+\varepsilon}\right] > 0 \\ \mathbf{n}^{\mathrm{T}}_{\partial\Omega_{ij}}\left(\mathbf{x}^{(0)}_{m+\varepsilon}\right) \cdot \left[\mathbf{x}^{(j)}_{m+\varepsilon} - \mathbf{x}^{(0)}_{m+\varepsilon}\right] < 0 \end{aligned}\right\} \text{for } \mathbf{n}_{\partial\Omega_{ij}} \to \Omega_i. \end{aligned} \tag{2.28}$$

The analytical conditions for such a non-passable flow of the first kind on $\partial\Omega_{ij}$ from domain Ω_i to Ω_j are given in Luo (2008, 2009, 2012), which can be stated in the following theorem.

Theorem 2.5 *For a discontinuous dynamical system in Eq. 2.1, there is a point* $\mathbf{x}^{(0)}(t_m) \equiv \mathbf{x}_m \in \partial\Omega_{ij}$ *at time* t_m *between two adjacent domains* Ω_α $(\alpha = i, j)$*. For an arbitrarily small* $\varepsilon > 0$*, there is a time interval* $(t_m, t_{m+\varepsilon}]$*. Suppose* $\mathbf{x}^{(\alpha)}(t_{m+}) = \mathbf{x}_m$ $(\alpha = i, j)$*. A flow* $\mathbf{x}^{(\alpha)}(t)$ *is* $C^{r_\alpha}_{(t_m, t_{m+\varepsilon}]}$*-continuous* $(r_\alpha \geq 1)$ *for time* t *with* $\left\|d^{r_\alpha+1}\mathbf{x}^{(\alpha)}/dt^{r_\alpha+1}\right\| < \infty$*. The flow* $\mathbf{x}^{(i)}(t)$ *and* $\mathbf{x}^{(j)}(t)$ *to the boundary* $\partial\Omega_{ij}$ *are non-passable of the second kind (or a source flow) if and only if*

$$\begin{array}{ll} \textit{either} & \left.\begin{array}{l} G^{(i)}_{\partial\Omega_{ij}}(\mathbf{x}_m, t_{m+}, \mathbf{p}_i, \boldsymbol{\lambda}) < 0 \\ G^{(j)}_{\partial\Omega_{ij}}(\mathbf{x}_m, t_{m+}, \mathbf{p}_j, \boldsymbol{\lambda}) > 0 \end{array}\right\} \textit{for } \mathbf{n}_{\partial\Omega_{ij}} \to \Omega_j, \\ \textit{or} & \left.\begin{array}{l} G^{(i)}_{\partial\Omega_{ij}}(\mathbf{x}_m, t_{m+}, \mathbf{p}_i, \boldsymbol{\lambda}) > 0 \\ G^{(j)}_{\partial\Omega_{ij}}(\mathbf{x}_m, t_{m+}, \mathbf{p}_j, \boldsymbol{\lambda}) < 0 \end{array}\right\} \textit{for } \mathbf{n}_{\partial\Omega_{ij}} \to \Omega_i. \end{array} \tag{2.29}$$

Proof See the proof as in Luo (2008, 2009, 2012). ■

From Luo (2008, 2009, 2012), the definition of the $(m_i : m_j)$ non-passable flow of the second kind is stated as follows.

Definition 2.14 For a discontinuous dynamic system in Eq. 2.1, there is a point $\mathbf{x}^{(0)}(t_m) \equiv \mathbf{x}_m \in \partial\Omega_{ij}$ at time t_m between two adjacent domains Ω_α $(\alpha = i, j)$. For an arbitrarily small $\varepsilon > 0$, there is a time interval $(t_m, t_{m+\varepsilon}]$. Suppose $\mathbf{x}^{(\alpha)}(t_{m+}) = \mathbf{x}_m$. A flow $\mathbf{x}^{(\alpha)}(t)$ is $C^{r_\alpha}_{(t_m, t_{m+\varepsilon}]}$-continuous $(r_\alpha \geq m_\alpha + 1, \alpha = i, j)$ for time t with $\left\| d^{r_\alpha+1}\mathbf{x}^{(\alpha)}/dt^{r_\alpha+1} \right\| < \infty$. The flow $\mathbf{x}^{(i)}(t)$ of the (m_i)th-order and $\mathbf{x}^{(j)}(t)$ of the (m_j)th-order to the boundary $\partial\Omega_{ij}$ are $(m_i : m_j)$ *-non-passable* of the second kind (or called an $(m_i : m_j)$-source flow) if

$$\left.\begin{array}{l} G^{(s_i,i)}_{\partial\Omega_{ij}}(\mathbf{x}_m, t_{m+}, \mathbf{p}_i, \boldsymbol{\lambda}) = 0 \ \text{ for } s_i = 0, 1, \cdots, m_i - 1, \\ G^{(2k_i,i)}_{\partial\Omega_{ij}}(\mathbf{x}_m, t_{m+}, \mathbf{p}_i, \boldsymbol{\lambda}) \neq 0, \end{array}\right\} \tag{2.30}$$

$$\left.\begin{array}{l} G^{(s_j,j)}_{\partial\Omega_{ij}}(\mathbf{x}_m, t_{m+}, \mathbf{p}_j, \boldsymbol{\lambda}) = 0 \ \text{ for } s_j = 0, 1, \cdots, m_j - 1, \\ G^{(2k_j,j)}_{\partial\Omega_{ij}}(\mathbf{x}_m, t_{m+}, \mathbf{p}_j, \boldsymbol{\lambda}) \neq 0, \end{array}\right\} \tag{2.31}$$

$$\begin{array}{ll} \text{either} & \left.\begin{array}{l} \mathbf{n}^{\mathrm{T}}_{\partial\Omega_{ij}}\left(\mathbf{x}^{(0)}_{m+\varepsilon}\right) \cdot \left[\mathbf{x}^{(i)}_{m+\varepsilon} - \mathbf{x}^{(0)}_{m+\varepsilon}\right] < 0 \\ \mathbf{n}^{\mathrm{T}}_{\partial\Omega_{ij}}\left(\mathbf{x}^{(0)}_{m+\varepsilon}\right) \cdot \left[\mathbf{x}^{(j)}_{m+\varepsilon} - \mathbf{x}^{(0)}_{m+\varepsilon}\right] > 0 \end{array}\right\} \text{for } \mathbf{n}_{\partial\Omega_{ij}} \to \Omega_j, \\ \text{or} & \left.\begin{array}{l} \mathbf{n}^{\mathrm{T}}_{\partial\Omega_{ij}}\left(\mathbf{x}^{(0)}_{m+\varepsilon}\right) \cdot \left[\mathbf{x}^{(i)}_{m+\varepsilon} - \mathbf{x}^{(0)}_{m+\varepsilon}\right] > 0 \\ \mathbf{n}^{\mathrm{T}}_{\partial\Omega_{ij}}\left(\mathbf{x}^{(0)}_{m+\varepsilon}\right) \cdot \left[\mathbf{x}^{(j)}_{m+\varepsilon} - \mathbf{x}^{(0)}_{m+\varepsilon}\right] < 0 \end{array}\right\} \text{for } \mathbf{n}_{\partial\Omega_{ij}} \to \Omega_i. \end{array} \tag{2.32}$$

The analytical conditions for such a $(m_i : m_j)$ non-passable flow of the second kind on $\partial\Omega_{ij}$ from domain Ω_i to Ω_j are given in Luo (Luo, 2009; Luo, 2012; Luo, 2005), which can be stated in the following theorem.

Theorem 2.6 *For a discontinuous dynamical system in Eq. 2.1, there is a point* $\mathbf{x}^{(0)}(t_m) \equiv \mathbf{x}_m \in \partial\Omega_{ij}$ *at time* t_m *between two adjacent domains* Ω_α $(\alpha = i, j)$. *For an arbitrarily small* $\varepsilon > 0$, *there is a time interval* $(t_m, t_{m+\varepsilon}]$. *Suppose* $\mathbf{x}^{(\alpha)}(t_{m+}) = \mathbf{x}_m$. *A flow* $\mathbf{x}^{(\alpha)}(t)$ *is* $C^{r_\alpha}_{(t_m, t_{m+\varepsilon}]}$*-continuous for time t with* $\left\| d^{r_\alpha+1}\mathbf{x}^{(\alpha)}/dt^{r_\alpha+1} \right\| < \infty (r_\alpha \geq m_\alpha + 1$, $\alpha = i, j)$. *The flow* $\mathbf{x}^{(i)}(t)$ *of the* (m_i)th*-order and* $\mathbf{x}^{(j)}(t)$ *of the* (m_j)th*-order to the boundary* $\partial\Omega_{ij}$ *are* $(m_i : m_j)$*-non-passable of the second kind (or* $(m_i : m_j)$*-source flow) if and only if*

$$G^{(s_i,i)}_{\partial\Omega_{ij}}(\mathbf{x}_m, t_{m+}, \mathbf{p}_i, \boldsymbol{\lambda}) = 0 \;\; for \;\; s_i = 0, 1, \cdots, m_i - 1; \tag{2.33}$$

$$G^{(s_j,j)}_{\partial\Omega_{ij}}(\mathbf{x}_m, t_{m+}, \mathbf{p}_j, \boldsymbol{\lambda}) = 0 \;\; for \;\; s_j = 0, 1, \cdots, m_j - 1; \tag{2.34}$$

$$\begin{aligned} & either \;\; \left. \begin{aligned} & G^{(2k_i,i)}_{\partial\Omega_{ij}}(\mathbf{x}_m, t_{m+}, \mathbf{p}_i, \boldsymbol{\lambda}) < 0 \\ & G^{(2k_j,j)}_{\partial\Omega_{ij}}(\mathbf{x}_m, t_{m+}, \mathbf{p}_j, \boldsymbol{\lambda}) > 0 \end{aligned} \right\} for \;\; \mathbf{n}_{\partial\Omega_{ij}} \to \Omega_j, \\ & or \;\; \left. \begin{aligned} & G^{(2k_i,i)}_{\partial\Omega_{ij}}(\mathbf{x}_m, t_{m+}, \mathbf{p}_i, \boldsymbol{\lambda}) > 0 \\ & G^{(2k_j,j)}_{\partial\Omega_{ij}}(\mathbf{x}_m, t_{m+}, \mathbf{p}_j, \boldsymbol{\lambda}) < 0 \end{aligned} \right\} for \;\; \mathbf{n}_{\partial\Omega_{ij}} \to \Omega_i. \end{aligned} \tag{2.35}$$

Proof See the proof as in Luo (2008, 2009, 2012). ■

As in Luo (2008, 2009, 2012), a tangential flow to the boundary $\partial\Omega_{ij}$ is presented as follows.

Definition 2.15 For a discontinuous dynamical system in Eq. 2.1, there is a point $\mathbf{x}^{(0)}(t_m) \equiv \mathbf{x}_m \in \partial\Omega_{ij}$ at time t_m between two adjacent domains Ω_α $(\alpha = i, j)$. Suppose $\mathbf{x}^{(\alpha)}(t_{m\pm}) = \mathbf{x}_m$ $(\alpha \in \{i, j\})$. For an arbitrarily small $\varepsilon > 0$, there is a time interval $[t_{m-\varepsilon}, t_{m+\varepsilon}]$. A flow $\mathbf{x}^{(\alpha)}(t)$ is $C^{r_\alpha}_{[t_{m-\varepsilon}, t_{m+\varepsilon}]}$-continuous $(r_\alpha \geq 2)$ for time t. The flow $\mathbf{x}^{(\alpha)}(t)$ in domain Ω_α is *tangential* to the boundary $\partial\Omega_{ij}$ if

$$G^{(0,\alpha)}_{\partial\Omega_{ij}}(\mathbf{x}_m, t_m, \mathbf{p}_\alpha, \boldsymbol{\lambda}) = 0 \text{ and } G^{(1,\alpha)}_{\partial\Omega_{ij}}(\mathbf{x}_m, t_m, \mathbf{p}_\alpha, \boldsymbol{\lambda}) \neq 0; \tag{2.36}$$

$$\begin{aligned} & \text{either} \;\; \left. \begin{aligned} & \mathbf{n}^{\mathrm{T}}_{\partial\Omega_{ij}}(\mathbf{x}^{(0)}_{m-\varepsilon}) \cdot [\mathbf{x}^{(0)}_{m-\varepsilon} - \mathbf{x}^{(\alpha)}_{m-\varepsilon}] > 0 \\ & \mathbf{n}^{\mathrm{T}}_{\partial\Omega_{ij}}(\mathbf{x}^{(0)}_{m+\varepsilon}) \cdot [\mathbf{x}^{(\alpha)}_{m+\varepsilon} - \mathbf{x}^{(0)}_{m+\varepsilon}] < 0 \end{aligned} \right\} \text{for } \mathbf{n}_{\partial\Omega_{ij}} \to \Omega_\beta, \\ & \text{or} \;\; \left. \begin{aligned} & \mathbf{n}^{\mathrm{T}}_{\partial\Omega_{ij}}(\mathbf{x}^{(0)}_{m-\varepsilon}) \cdot [\mathbf{x}^{(0)}_{m-\varepsilon} - \mathbf{x}^{(\alpha)}_{m-\varepsilon}] < 0 \\ & \mathbf{n}^{\mathrm{T}}_{\partial\Omega_{ij}}(\mathbf{x}^{(0)}_{m+\varepsilon}) \cdot [\mathbf{x}^{(\alpha)}_{m+\varepsilon} - \mathbf{x}^{(0)}_{m+\varepsilon}] > 0 \end{aligned} \right\} \text{for } \mathbf{n}_{\partial\Omega_{ij}} \to \Omega_\alpha. \end{aligned} \tag{2.37}$$

As in Luo (2008, 2009, 2012), the conditions for the tangential flow (grazing flow) are given by the following theorem.

Theorem 2.7 *For a discontinuous dynamical system in Eq. 2.1, there is a point* $\mathbf{x}^{(0)}(t_m) \equiv \mathbf{x}_m \in \partial\Omega_{ij}$ *at time* t_m *between two adjacent domains* $\Omega_\alpha(\alpha = i,j)$. *Suppose* $\mathbf{x}^{(\alpha)}(t_{m\pm}) = \mathbf{x}_m\ (\alpha \in \{i,j\})$. *For an arbitrarily small* $\varepsilon > 0$, *there is a time interval* $[t_{m-\varepsilon}, t_{m+\varepsilon}]$. *A flow* $\mathbf{x}^{(\alpha)}(t)$ *is* $C^r_{[t_{m-\varepsilon},t_{m+\varepsilon}]}$*-continuous* $(r \geq 2)$ *for time t with* $\left\| d^{r_\alpha+1}\mathbf{x}^{(\alpha)}/dt^{r_\alpha+1} \right\| < \infty$. *The flow* $\mathbf{x}^{(\alpha)}(t)$ *in domain* Ω_α *is tangential to the boundary* $\partial\Omega_{ij}$ *if and only if*

$$G^{(0,\alpha)}_{\partial\Omega_{ij}}(\mathbf{x}_m, t_m, \mathbf{p}_\alpha, \boldsymbol{\lambda}) = 0 \quad for\ \alpha \in \{i,j\}; \tag{2.38}$$

$$\begin{array}{ll} either & G^{(1,\alpha)}_{\partial\Omega_{ij}}(\mathbf{x}_m, t_m, \mathbf{p}_\alpha, \boldsymbol{\lambda}) < 0 \ for\ \mathbf{n}_{\partial\Omega_{ij}} \to \Omega_\beta, \\ or & G^{(1,\alpha)}_{\partial\Omega_{ij}}(\mathbf{x}_m, t_m, \mathbf{p}_\alpha, \boldsymbol{\lambda}) > 0 \ for\ \mathbf{n}_{\partial\Omega_{ij}} \to \Omega_\alpha. \end{array} \tag{2.39}$$

Proof See the proof as in Luo (Luo, 2009; Luo, 2012; Luo, 2008). ■

From Luo (2008, 2009, 2012), a $(2k_\alpha - 1)$th-order tangential (grazing) flow to the boundary $\partial\Omega_{ij}$ is presented as follows.

Definition 2.16 For a discontinuous dynamical system in Eq. 2.1, there is a point $\mathbf{x}^{(0)}(t_m) \equiv \mathbf{x}_m \in \partial\Omega_{ij}$ at time t_m between two adjacent domains $\Omega_\alpha\ (\alpha = i, j)$. Suppose $\mathbf{x}^{(\alpha)}(t_{m\pm}) = \mathbf{x}_m\ (\alpha \in \{i,j\})$. For an arbitrarily small $\varepsilon > 0$, there is a time interval $[t_{m-\varepsilon}, t_{m+\varepsilon}]$. A flow $\mathbf{x}^{(\alpha)}(t)$ is $C^{r_\alpha}_{[t_{m-\varepsilon},t_{m+\varepsilon}]}$-continuous $(r_\alpha \geq k_\alpha + 1)$ with $\left\| d^{r_\alpha+1}\mathbf{x}^{(\alpha)}/dt^{r_\alpha+1} \right\| < \infty$ for time t. A flow $\mathbf{x}^{(\alpha)}(t)$ in Ω_α is *tangential* to the boundary $\partial\Omega_{ij}$ of the $(2k_\alpha - 1)$th-order if

$$\begin{array}{l} G^{(s_\alpha,\alpha)}_{\partial\Omega_{ij}}(\mathbf{x}_m, t_m, \mathbf{p}_\alpha, \boldsymbol{\lambda}) = 0 \text{ for } s_\alpha = 0, 1, \cdots, 2k_\alpha - 2; \\ G^{(2k_\alpha-1,\alpha)}_{\partial\Omega_{ij}}(\mathbf{x}_m, t_m, \mathbf{p}_\alpha, \boldsymbol{\lambda}) \neq 0; \end{array} \tag{2.40}$$

$$\begin{array}{ll} \text{either} & \left.\begin{array}{l} \mathbf{n}^{\mathrm{T}}_{\partial\Omega_{ij}}(\mathbf{x}^{(0)}_{m-\varepsilon}) \cdot [\mathbf{x}^{(0)}_{m-\varepsilon} - \mathbf{x}^{(\alpha)}_{m-\varepsilon}] > 0 \\ \mathbf{n}^{\mathrm{T}}_{\partial\Omega_{ij}}\left(\mathbf{x}^{(0)}_{m+\varepsilon}\right) \cdot \left[\mathbf{x}^{(\alpha)}_{m+\varepsilon} - \mathbf{x}^{(0)}_{m+\varepsilon}\right] < 0 \end{array}\right\} \text{for } \mathbf{n}_{\partial\Omega_{ij}} \to \Omega_\beta, \\ \text{or} & \left.\begin{array}{l} \mathbf{n}^{\mathrm{T}}_{\partial\Omega_{ij}}(\mathbf{x}^{(0)}_{m-\varepsilon}) \cdot [\mathbf{x}^{(0)}_{m-\varepsilon} - \mathbf{x}^{(\alpha)}_{m-\varepsilon}] < 0 \\ \mathbf{n}^{\mathrm{T}}_{\partial\Omega_{ij}}\left(\mathbf{x}^{(0)}_{m+\varepsilon}\right) \cdot \left[\mathbf{x}^{(\alpha)}_{m+\varepsilon} - \mathbf{x}^{(0)}_{m+\varepsilon}\right] > 0 \end{array}\right\} \text{for } \mathbf{n}_{\partial\Omega_{ij}} \to \Omega_\alpha. \end{array} \tag{2.41}$$

As in Luo (2008, 2009, 2012), the analytical conditions for the $(2k_\alpha - 1)$th-order tangential flow (grazing flow) are given by the following theorem.

Theorem 2.8 *For a discontinuous dynamical system in Eq. 2.1, there is a point* $\mathbf{x}^{(0)}(t_m) \equiv \mathbf{x}_m \in \partial\Omega_{ij}$ *at time* t_m *between two adjacent domains* Ω_α $(\alpha = i, j)$. *Suppose* $\mathbf{x}^{(\alpha)}(t_{m\pm}) = \mathbf{x}_m$ $(\alpha \in \{i, j\})$. *For an arbitrarily small* $\varepsilon > 0$, *there is a time interval* $[t_{m-\varepsilon}, t_{m+\varepsilon}]$. *A flow* $\mathbf{x}^{(\alpha)}(t)$ *is* $C^{r_\alpha}_{[t_{m-\varepsilon}, t_{m+\varepsilon}]}$*-continuous* $(r_\alpha \geq k_\alpha + 1)$ *for time t with* $\left\| d^{r_\alpha+1}\mathbf{x}^{(\alpha)}/dt^{r_\alpha+1} \right\| < \infty$. *A flow* $\mathbf{x}^{(\alpha)}(t)$ *in* Ω_α *is tangential to the boundary* $\partial\Omega_{ij}$ *of the* $(2k_\alpha - 1)$th-*order if and only if*

$$G^{(s_\alpha,\alpha)}_{\partial\Omega_{ij}}(\mathbf{x}_m, t_m, \mathbf{p}_\alpha, \boldsymbol{\lambda}) = 0 \textit{ for } s_\alpha = 0, 1, \cdots, 2k_\alpha - 2; \tag{2.42}$$

$$\begin{aligned} &\textit{either} \quad G^{(2k_\alpha - 1,\alpha)}_{\partial\Omega_{ij}}(\mathbf{x}_m, t_m, \mathbf{p}_\alpha, \boldsymbol{\lambda}) < 0 \text{ for } \mathbf{n}_{\partial\Omega_{ij}} \to \Omega_\beta, \\ &\textit{or} \quad G^{(2k_\alpha - 1,\alpha)}_{\partial\Omega_{ij}}(\mathbf{x}_m, t_m, \mathbf{p}_\alpha, \boldsymbol{\lambda}) > 0 \text{ for } \mathbf{n}_{\partial\Omega_{ij}} \to \Omega_\alpha. \end{aligned} \tag{2.43}$$

Proof See the proof as in Luo (2008, 2009, 2012). ■

References

Luo, A. C. J. (2005). A theory for non-smooth dynamic systems on the connectable domains. *Communications in Nonlinear Science and Numerical Simulation, 10*, 1–55.

Luo, A. C. J. (2006). *Singularity and dynamics on discontinuous vector fields*. Elsevier.

Luo, A. C. J. (2008). A theory for flow switchability in discontinuous dynamical systems. *Nonlinear Analysis: Hybrid Systems, 2*(4), 1030–1061.

Luo, A. C. J. (2009). *Discontinuous dynamical systems on time-varying domains*. Higher Education Press/Springer.

Luo, A. C. J. (2012). *Discontinuous dynamical systems*. Higher Education Press/Springer.

Chapter 3
A Duffing-Like Oscillator with Signum Nonlinearity

Contents

3.1 Discontinuous Dynamical System 23
3.2 Analytical Switching Conditions 25
3.3 Mapping Structures 27
3.4 Bifurcation Scenarios 29
3.5 Parameter Maps 31
3.6 Simulation Illustrations 32
References 34

In this chapter, the nonlinear dynamics of a Duffing-like system with signum function are studied from the theory of discontinuous dynamical systems. The necessary and sufficient conditions for motion switchability on the boundary between two domains are presented to understand the switching mechanism. The switching velocity with varying different system parameters and the parameter mappings are carried out to illustrate the dynamical motions. The attraction basins are depicted to express the coexistence of the Duffing-like oscillator with different initial values, and the coexisting trajectories in phase-space with various initial conditions are also exhibited.

3.1 Discontinuous Dynamical System

As in Min and Chen (2020), consider a periodically forced Duffing-like system with signum nonlinearity:

$$\ddot{x} + a\dot{x} - x + b\,\mathrm{sgn}(x) = Q\sin(\omega t) \tag{3.1}$$

F. Min, *Discontinuous Dynamics and System Synchronization*, Nonlinear Systems and Complexity 39, https://doi.org/10.1007/978-3-031-66648-3_3

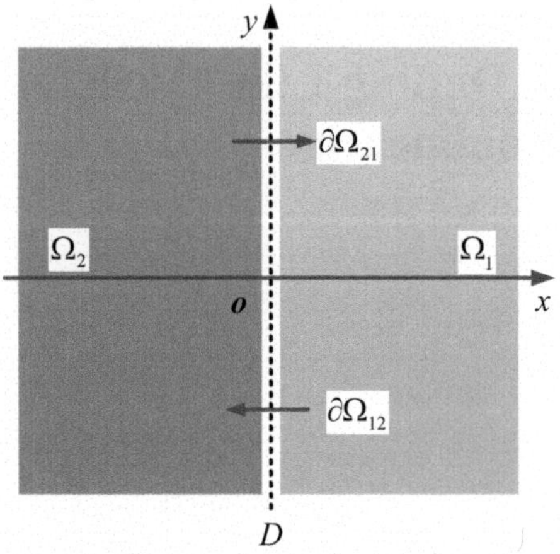

Fig. 3.1 Domain partitions in phase plane for constant displacement D

where a is the damping coefficient of the Duffing-like system and Q and ω are the excitation amplitude and frequency, respectively. As there is a signum function in Eq. 3.1, this system is a non-smooth system, existing switching motions on the separation boundary, as shown in Fig. 3.1.

In Fig. 3.1, here are two domains, $\Omega_\alpha(\alpha \in \{1,2\})$, and a separation boundary D labeled by a straight dotted line. The motion in each domain can be described by an independent continuous dynamical system. The possible directions of the switching flow are indicated by the arrows crossing the boundary.

In phase plane, the state variables and the vector fields are described as

$$\mathbf{x} = (x, \dot{x})^T = (x, y)^T \text{ and } \mathbf{F} = (y, F)^T. \tag{3.2}$$

where F is the scalar force.

Two domains and the separation boundary of the system are given by

$$\left.\begin{aligned} \Omega_1 &= \{(x,y)|x \in (D, +\infty)\}, \\ \Omega_2 &= \{(x,y)|x \in (-\infty, D)\}, \end{aligned}\right\} \tag{3.3}$$

and

$$\partial\Omega_{\alpha\beta} = \overline{\Omega}_\alpha \cap \overline{\Omega}_\beta = \{(x,y)|\varphi_{\alpha\beta}(x,y) \equiv x - D = 0\} \tag{3.4}$$

It should be noted that $\varphi_{\alpha\beta}(x,y)$ stands for the boundary from Ω_α to $\Omega_\beta(\alpha, \beta \in \{1,2\}$ and $\alpha \neq \beta)$ and D belongs to real numbers. From Luo (2005, 2008, 2009, 2012), the system equation in Eq. 3.1 can be expressed by

$$\dot{\mathbf{x}} = \mathbf{F}_{\lambda}^{(\kappa)}(\mathbf{x}, t), (\kappa, \lambda \in \{0, 1, 2\}) \tag{3.5}$$

where

$$\left.\begin{aligned}
&\mathbf{F}_{\alpha}^{(\alpha)}(\mathbf{x}, t) = \left(y, F_{\alpha}(\mathbf{x}, t)^{T}\right) \text{ in } \Omega_{\alpha}(\alpha \in \{1, 2\}),\\
&\mathbf{F}_{\alpha}^{(\beta)}(\mathbf{x}, t) = \left(y, F_{\beta}(\mathbf{x}, t)^{T}\right) \text{ in } \Omega_{\alpha}(\alpha \neq \beta \in \{1, 2\});\\
&\mathbf{F}_{0}^{(0)}(\mathbf{x}, t) = (y, 0)^{T} \text{ on } \partial\Omega_{\alpha\beta} \text{ for stick,}\\
&\mathbf{F}_{0}^{(0)}(\mathbf{x}, t) \in \left[\mathbf{F}_{\alpha}^{(\alpha)}(\mathbf{x}, t), \mathbf{F}_{\beta}^{(\beta)}(\mathbf{x}, t)\right] \text{ on } \partial\Omega_{\alpha\beta} \text{ for non-stick}
\end{aligned}\right\} \tag{3.6}$$

$$F_{\alpha}(\mathbf{x}, t) = -a^{(\alpha)}y + x + (-b)^{(\alpha)} + Q\sin(\omega t) \quad (\alpha = 1, 2). \tag{3.7}$$

The superscript κ and subscript λ with nonzero values represent two domains for α, $\beta \in \{1, 2\}$. $\mathbf{F}_{\alpha}^{(\alpha)}(\mathbf{x}, t)$ is the real vector field in the α-domain—$\mathbf{F}_{\alpha}^{(\beta)}(\mathbf{x}, t)$ is the imaginary vector field in the α-domain, which is decided by the vector field in the β-domain. The vector field on the boundary is expressed as $\mathbf{F}_{0}^{(0)}(\mathbf{x}, t)$. $F_{\alpha}(\mathbf{x}, t)$ is the scalar force in α-domain, and the domain forces in Eq. 3.1 are denoted by Eq. 3.7, where the forces include the linear spring forces, the linear damping forces, and the excitation force.

3.2 Analytical Switching Conditions

From Luo (2005, 2008, 2009, 2012), the necessary and sufficient conditions for system motions in Eq. 3.1, passing the boundary $\partial\Omega_{\alpha\beta}$ with the corresponding normal vector $\mathbf{n}_{\partial\Omega_{\alpha\beta}}$ pointing to domain Ω_{α}, are given by

$$\left.\begin{aligned}
&\left.\begin{aligned}
G^{(0,\alpha)}(\mathbf{x}_m, t_{m-}) = \mathbf{n}_{\partial\Omega_{\alpha\beta}}^{T} \cdot \mathbf{F}^{(\alpha)}(\mathbf{x}_m, t_{m-}) < 0,\\
G^{(0,\beta)}(\mathbf{x}_m, t_{m+}) = \mathbf{n}_{\partial\Omega_{\alpha\beta}}^{T} \cdot \mathbf{F}^{(\beta)}(\mathbf{x}_m, t_{m+}) < 0
\end{aligned}\right\} \text{from } \Omega_{\alpha} \to \Omega_{\beta};\\
&\left.\begin{aligned}
G^{(0,\beta)}(\mathbf{x}_m, t_{m-}) = \mathbf{n}_{\partial\Omega_{\alpha\beta}}^{T} \cdot \mathbf{F}^{(\beta)}(\mathbf{x}_m, t_{m-}) > 0,\\
G^{(0,\alpha)}(\mathbf{x}_m, t_{m+}) = \mathbf{n}_{\partial\Omega_{\alpha\beta}}^{T} \cdot \mathbf{F}^{(\alpha)}(\mathbf{x}_m, t_{m+}) > 0,
\end{aligned}\right\} \text{from } \Omega_{\beta} \to \Omega_{\alpha}
\end{aligned}\right\} \tag{3.8}$$

for $\mathbf{n}_{\partial\Omega_{\alpha\beta}} \to \Omega_{\alpha}$,
where α, $\beta \in \{1, 2\}$ and $\alpha \neq \beta$ with

$$\mathbf{n}_{\partial\Omega_{\alpha\beta}} = \nabla\varphi_{\alpha\beta} = \left(\partial_{x}\varphi_{\alpha\beta}, \partial_{y}\varphi_{\alpha\beta}\right)^{T}_{(x_m, y_m)} \tag{3.9}$$

Note that $\nabla = (\partial_x, \partial_y)^T$ is the Hamilton operator with $\partial_x(\cdot) = \partial(\cdot)/\partial x$ and $\partial_y(\cdot) = \partial(\cdot)/\partial y$. $\mathbf{F}^{(\alpha)}(\mathbf{x}, t)$ is the real vector field in α-domain and imaginary in

β-domain. $\mathbf{F}^{(\beta)}(\mathbf{x}, t)$ is the real vector field in β-domain and imaginary in α-domain. The switching time t_m is the system motions on the boundary, and the switching time $t_{m\pm} = t_m \pm 0$ represents motions in domains very close to the separation boundary other than on the boundary, i.e., the switching time t_{m-} and t_{m+} indicate the time of the system trajectories just approaches or leaves the separation boundary, respectively.

The necessary and sufficient conditions of a grazing motion to the boundary $\partial\Omega_{\alpha\beta}$ are given by

$$\left.\begin{aligned} G^{(0,\alpha)}(\mathbf{x}_m, t_{m\pm}) &= \mathbf{n}^T_{\partial\Omega_{\alpha\beta}} \cdot \mathbf{F}^{(\alpha)}(\mathbf{x}_m, t_{m\pm}) = 0, \\ G^{(1,\alpha)}(\mathbf{x}_m, t_{m\pm}) &= \mathbf{n}^T_{\partial\Omega_{\alpha\beta}} \cdot D\mathbf{F}^{(\alpha)}(\mathbf{x}_m, t_{m\pm}) > 0 \end{aligned}\right\} \text{ in domain } \Omega_\alpha \tag{3.10}$$

for $\mathbf{n}_{\partial\Omega_{\alpha\beta}} \to \Omega_\alpha$

and

$$\left.\begin{aligned} G^{(0,\beta)}(\mathbf{x}_m, t_{m\pm}) &= \mathbf{n}^T_{\partial\Omega_{\alpha\beta}} \cdot \mathbf{F}^{(\beta)}(\mathbf{x}_m, t_{m\pm}) = 0, \\ G^{(1,\beta)}(\mathbf{x}_m, t_{m\pm}) &= \mathbf{n}^T_{\partial\Omega_{\alpha\beta}} \cdot D\mathbf{F}^{(\beta)}(\mathbf{x}_m, t_{m\pm}) < 0 \end{aligned}\right\} \text{ in domain } \Omega_\beta \tag{3.11}$$

for $\mathbf{n}_{\partial\Omega_{\alpha\beta}} \to \Omega_\alpha$

where

$$D\mathbf{F}^{(\alpha)}(\mathbf{x}, t) = \left(F_\alpha(\mathbf{x}, t), \nabla F^{(\alpha)}(\mathbf{x}, t) \cdot \mathbf{F}^{(\alpha)}(\mathbf{x}, t) + \partial_t F_\alpha(\mathbf{x}, t)\right)^T \tag{3.12}$$

According to Eqs. 3.4 and 3.9, the normal vector can be obtained as follows:

$$\mathbf{n}_{\partial\Omega_{12}} = \mathbf{n}_{\partial\Omega_{21}} = (1, 0)^T \tag{3.13}$$

From Eq. 3.13, the normal vector of the displacement boundary always points to domain Ω_1.

The zero and first order G-functions for $\alpha \in \{1, 2\}$ are described as

$$\left.\begin{aligned} G^{(0,\alpha)}(\mathbf{x}_m, t_{m\pm}) &= \mathbf{n}^T_{\partial\Omega_{\alpha\beta}} \cdot \mathbf{F}^{(\alpha)}(\mathbf{x}_m, t_{m\pm}) = y_m, \\ G^{(1,\alpha)}(\mathbf{x}_m, t_{m\pm}) &= \mathbf{n}^T_{\partial\Omega_{\alpha\beta}} \cdot D\mathbf{F}^{(\alpha)}(\mathbf{x}_m, t_{m\pm}) = F_\alpha(\mathbf{x}_m, t_{m\pm}), \end{aligned}\right\} \tag{3.14}$$

and

$$\left.\begin{aligned} G^{(1,1)}(\mathbf{x}_m, t_m) &= F_1(\mathbf{x}_m, t_{m\pm}) \\ G^{(1,2)}(\mathbf{x}_m, t_m) &= F_2(\mathbf{x}_m, t_{m\pm}) \end{aligned}\right\} \tag{3.15}$$

From Eqs. 3.8 and 3.14, the passable conditions of switching motions through the boundary are shown as

$$\left.\begin{aligned} G^{(0,1)}(\mathbf{x}_m, t_{m-}) &= y_{m-} < 0, \\ G^{(0,2)}(\mathbf{x}_m, t_{m+}) &= y_{m+} < 0 \end{aligned}\right\} \text{from } \Omega_1 \to \Omega_2 \tag{3.16}$$

$$\left.\begin{aligned} G^{(0,1)}(\mathbf{x}_m, t_{m-}) &= y_{m-} > 0, \\ G^{(0,2)}(\mathbf{x}_m, t_{m+}) &= y_{m+} > 0 \end{aligned}\right\} \text{from } \Omega_2 \to \Omega_1. \tag{3.17}$$

According to Eqs. 3.10 and 3.11, the condition of the grazing motion to the boundary is given by

$$\left.\begin{aligned} G^{(0,1)}(\mathbf{x}_m, t_{m\pm}) &= y_{m\pm} = 0, \\ G^{(1,1)}(\mathbf{x}_m, t_{m\pm}) &= F_1(x_m, t_{m\pm}) > 0 \end{aligned}\right\} \text{in domain } \Omega_1 \tag{3.18}$$

and

$$\left.\begin{aligned} G^{(0,2)}(\mathbf{x}_m, t_{m\pm}) &= y_{m\pm} = 0, \\ G^{(1,2)}(\mathbf{x}_m, t_{m\pm}) &= F_2(\mathbf{x}_m, t_{m\pm}) < 0 \end{aligned}\right\} \text{in domain } \Omega_2. \tag{3.19}$$

3.3 Mapping Structures

To illustrate the complexity motions of system, the mapping structures for the periodic motions are depicted in Fig. 3.2, where the system trajectories in Ω_α, starting and ending at the boundary (i.e., from $\Omega_{\alpha\beta}$ to $\Omega_{\beta\alpha}$), are sketched. The starting and ending points for mappings P_α in Ω_α are (D, y_k, t_k) and (D, y_{k+1}, t_{k+1}), respectively. According to the displacement boundary, define the switching sets as

$$\left.\begin{aligned} \Xi^+ &= \{(y_k, \omega t_k) | x_k = D^+\}, \\ \Xi^- &= \{(y_k, \omega t_k) | x_k = D^-\} \end{aligned}\right\} \tag{3.20}$$

The superscript "+" and "-" represents that the switching set approaches to the right and left side of the boundary, respectively. In Fig. 3.2, the dash lines infinitesimally close to the displacement boundary mean the switching sets Ξ. Therefore, two mappings are defined as

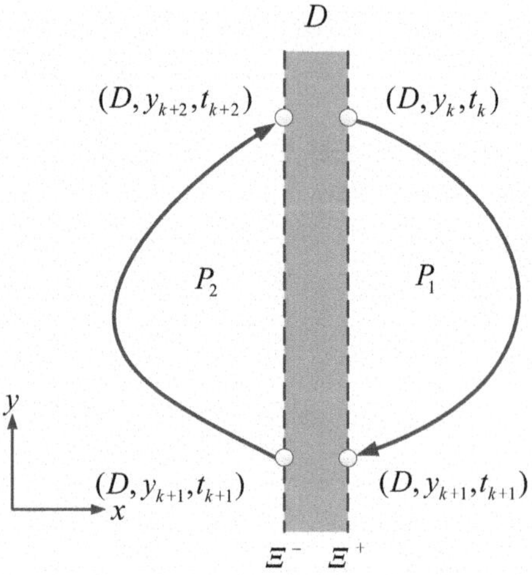

Fig. 3.2 Switching sets and mappings

$$P_1 : \Xi^+ \to \Xi^+, \quad P_2 : \Xi^- \to \Xi^- \tag{3.21}$$

Substitute Eqs. 3.20 into 3.21, and then the mappings can be expressed as

$$\left.\begin{array}{l} P_1 : (D^+, y_k, \omega t_k) \to \left(D^+, y_{k+1}, \omega t_{k+1}\right) \\ P_2 : (D^-, y_k, \omega t_k) \to \left(D^-, y_{k+1}, \omega t_{k+1}\right) \end{array}\right\} \tag{3.22}$$

A mapping structure for a periodic behavior with nonstick is described as

$$P = \underbrace{\left(P_2^{(k_{m2})} \circ P_1^{(k_{m1})}\right) \circ \cdots \circ \left(P_2^{(k_{12})} \circ P_1^{(k_{11})}\right)}_{m-terms} \tag{3.23}$$

where $k_{l\lambda} \in \{0, 1\}$ for $k_{l\lambda} \in \{0, 1\}$ and $\lambda \in \{1, 2\}$. $P_\lambda^{(0)} = 1$ and $P_\lambda^{(k)} = P_\lambda \circ P_\lambda^{(k-1)}$. For convenience, the mapping notations in Eq. 3.23 can be expressed as

$$P = \underbrace{P_{\left(2^{k_{m2}} 1^{k_{m1}}\right)} \cdots P_{\left(2^{k_{12}} 1^{k_{11}}\right)}}_{m-terms}. \tag{3.24}$$

Set $m = k_{12} = k_{11} = 1$ in Eq. 3.23—a mapping structure for one of the simplest, nonstick periodic motions passing through the switching boundary is obtained, which is presented in Fig. 3.2. The mapping structures of such a simple periodic motion is expressed by

$$P_{21} = P_2 \circ P_1 : \Xi^+ \to \Xi^- \tag{3.25}$$

Consider the above relations—the two mapping structures are

$$\left.\begin{array}{l}P_1 : (D^+, y_k, t_k) \rightarrow (D^+, y_{k+1}, t_{k+1}) \\ P_2 : (D^-, y_{k+1}, t_{k+1}) \rightarrow (D^-, y_{k+2}, t_{k+2})\end{array}\right\} \tag{3.26}$$

Without sliding in Eq. 3.1, the displacement boundary $D^+ = D^- = D$ exists. For $\mathbf{x}_{k+2} = P\mathbf{x}_k$, where $\mathbf{x}_k = (y_k, \omega t_k)^T$ during N-periods of excitation, the periodicity of the periodic motion is given by

$$y_{k+2} = y_k, \omega t_{k+2} = \omega t_k + 2N\pi. \tag{3.27}$$

From the Duffing-like oscillator in Eq. 3.1, the governing equations for the simplest periodic motion in Fig. 3.2 are described as

$$\left.\begin{array}{l}f_1^{(1)}(y_k, \omega t_k, y_{k+1}, \omega t_{k+1}) = 0, \\ f_2^{(1)}(y_k, \omega t_k, y_{k+1}, \omega t_{k+1}) = 0; \\ f_1^{(2)}(y_{k+1}, \omega t_k, y_{k+2}, \omega t_{k+2}) = 0, \\ f_2^{(2)}(y_{k+1}, \omega t_k, y_{k+2}, \omega t_{k+2}) = 0.\end{array}\right\} \tag{3.28}$$

In a similar way, the periodic motion for the generalized mapping structures can be described. Here, the numerical algorithm will be applied to obtain the initial switching set for periodic motions.

3.4 Bifurcation Scenarios

The system dynamic behaviors are investigated via the numerical scheme, with the domains depicted by parameter mappings. The computational error tolerance is given by 10^{-8}. Behaviors on the boundary are illustrated by bifurcation diagrams, related to parameters such as a, b, and Q, as shown in Fig. 3.3. The coexistence motions of the Duffing-like system and the corresponding ranges of each parameter are also listed in Table 3.1.

First, let the system parameters $b = 1$, $Q = 1.1$ and $\omega = 1$—the bifurcation diagrams with decreasing damping coefficient a from 1.45 to 1.03 at a step size of $\Delta a = 0.0005$ are plotted with red and blue circles for initial conditions $x_0 = 0.1$, $y_0 = 0.1$, $t_0 = 0$ and $x_0 = 0.002$, $y_0 = -0.001$, $t_0 = 0$, respectively, as shown in Fig. 3.3a, where the acronyms "PD" and "PF" represent "period-doubling bifurcation" and "pitchfork bifurcation," respectively. The mapping structures (i.e., P_{21}) have been marked. With the damping coefficient $a \in (1.04, 1.43)$, the system trajectories change from chaos to periodic motion, via some inverse period-doubling bifurcations at $a = 1.11$ and $a = 1.14$. The coexisting bifurcations illustrate the properties of transition and complementarity. If $a < 1.04$, the system trajectories are diverged.

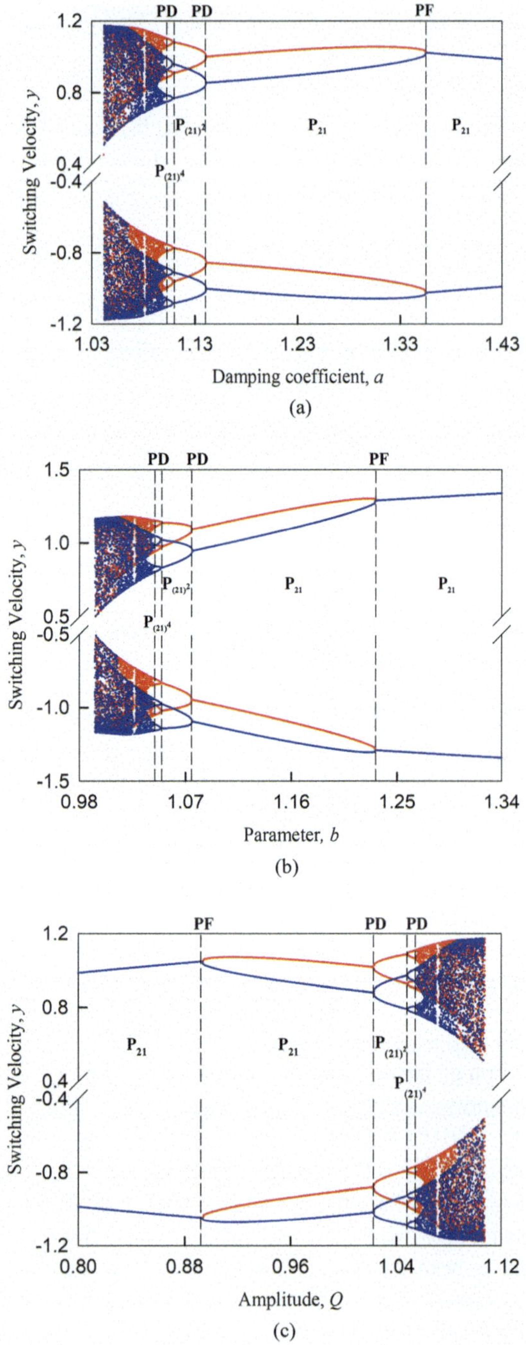

Fig. 3.3 Bifurcation diagrams of switching velocity y on the boundary with (**a**) $a \in [1.03, 1.45]$, (**b**) $b \in [0.98, 1.34]$, (**c**) $Q \in (0.8, 1.12)$

Table 3.1 Coexistence mapping structures of the ranges parameter a, b, and Q

Coexistence mapping structures	Ranges of a ($b = 1$, $Q = 1.1$, $\omega = 1$)	Ranges of b ($a = 1.05$, $Q = 1.1$, $\omega = 1$)	Ranges of Q ($a = 1.05$, $b = 1$, $\omega = 1$)
P_{21} with P_{21}	(1.14, 1.45)	(1.075, 1.34)	(0.80, 1.0226)
$P_{(21)^2}$ with $P_{(21)^2}$	(1.11, 1.14)	(1.05, 1.075)	(1.0226, 1.048)
$P_{(21)^4}$ with $P_{(21)^4}$	(1.103, 1.11)	(1.044, 1.05)	(1.048, 1.054)
Chaos with chaos	(1.04,1.103)	(0.99, 1.044)	(1.054, 1.108)

Next, with the damping coefficient $a = 1.05$, amplitude $Q = 1.1$, and frequency $\omega = 1$, the global bifurcation diagrams by varying b are presented in Fig. 3.3b, where the system lies in periodic motions with parameter $b \in (1.044, 1.34)$. With decreasing b from 1.34, a pitchfork bifurcation occurs at $b = 1.232$, and two limit cycles coexist until a period-doubling bifurcation appears at $b = 1.075$, and then the periodic orbit with mapping structures $P_{(21)^2}$ shows up until another period-doubling bifurcation occurs at $b = 1.05$. While b decreases, the motion direction is similar to that of damping coefficient decreasing, which is from period to chaos and eventually diverges.

Moreover, let $a = 1.05$, $b = 1$, $\omega = 1$, and $Q \in (0.8, 1.12)$, with the amplitude Q at steps of $\Delta Q = 0.0005$, the bifurcation diagrams of coexisting switching velocity y with amplitude Q are calculated as shown in Fig. 3.3c, where a pitchfork bifurcation occurs at about $Q = 0.89$, and two attractors are denoted in red and blue, respectively. By increasing amplitude Q in Fig. 3.3c, the trajectories of the system get into chaos from periodic motion through a set of period-doubling bifurcation at about $Q = 1.0226$, $Q = 1.048$, and $Q = 1.054$, in which the different coexisting behaviors with the same mapping structures are also observed. After $Q = 1.108$, the trajectory diverges.

3.5 Parameter Maps

For a global view of the dynamical behaviors of the system in Eq. 3.1, the parameter mappings are obtained from the analytical conditions, as shown in Fig. 3.4a with $Q = 1.1$, $\omega = 1$, and Fig. 3.4b with $b = 1$, $\omega = 1$, where the mapping structures P_{21}, $P_{(21)^2}$, $P_{(21)^3}$, and $P_{(21)^4}$ are marked by blue, green, dark blue, and yellow regions and red domains represent multiple periodic and chaotic motions. From Fig. 3.4a, with decreasing a and b, the mapping structures of the system will change from P_{21} to $P_{(21)^2}$ through a brief chaotic region at first, then turn to $P_{(21)^4}$, and finally get into chaos with mapping structure of $P_{(21)^3}$ among it. From Fig. 3.6b, the mapping structures become more and more complex with increasing amplitude Q and decreasing damping coefficient a, and this motion pattern is similar with that in Fig. 3.4a, which is consistent with that of the bifurcation diagrams in Fig. 3.3.

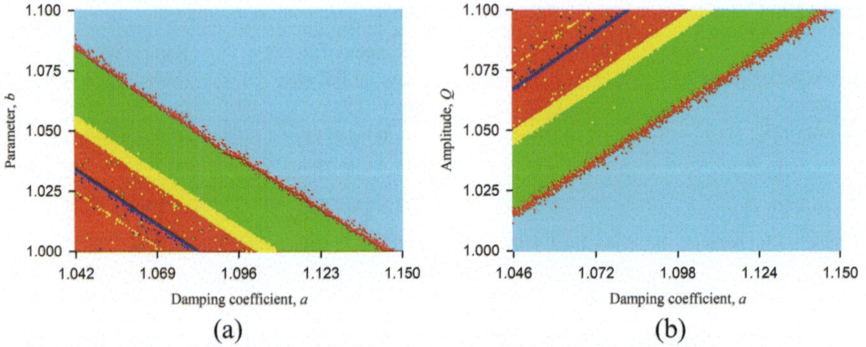

Fig. 3.4 Parameter mappings: (**a**) a-b, (**b**) a-Q (blue: P_{21}; green: $P_{(21)^2}$; dark blue: $P_{(21)^3}$; yellow: $P_{(21)^4}$; red: complex motions)

3.6 Simulation Illustrations

For a better understanding of motion switchability at the displacement boundary, a simple periodic motion and a chaotic motion of the nonlinear system in Eq. 3.1 will be illustrated. The symplectic algorithm is also used for numerical simulations of motions in the non-smooth Duffing-like oscillator with the same tolerance.

From the bifurcation diagrams in Fig. 3.3c and the parameter map in Fig. 3.4b, for $(a, b, Q, \omega) = (1.05, 1.0, 1.01, 1.0)$, the coexisting periodic motion with mapping structure P_{21} with different initial conditions can be observed. In Fig. 3.5a, b, the coexisting antisymmetric trajectories in the phase plane with initial values $(t_0, x_0, y_0) = (0.0, -1.007, -0.3248)$ and $(0.0, -0.2901, -0.4689)$ are plotted, respectively. The vertical dashed line $D = 0$ is the separation displacement boundary, and the periodicity of the periodic motion P_{21} can be observed clearly. The switching points are marked by red and green filled circular symbols, which represent the mapping structures of P_2 switching to P_1 and mapping of P_1 switching to P_2, respectively. In Fig. 3.5c, d, the time histories of displacement of the coexisting periodic motions are depicted, in which the curves are antisymmetric relative to the boundary. The corresponding mappings are labeled, and the mapping structures of the periodic orbits are clearly seen, in which the same periods T are also marked. In Fig. 3.5e, f, the time histories of G-functions are presented to illustrate the flow switchability, respectively. According to Eq. 3.14, the G-functions always satisfied $\mathrm{G}^{(0,2)} = G^{(0,1)}$, and on the separation boundary of two domains, the switching points have the same signs of G-functions. From the passible motion conditions in Eqs. 3.16 and 3.17, the coexisting periodic motions at the displacement boundary are always switchable, and the corresponding G-functions can also be switched, i.e., $\mathrm{G}^{(0,2)} = G^{(0,1)} > 0.0$ or $\mathrm{G}^{(0,2)} = G^{(0,1)} < 0.0$.

In addition to the periodic motions, chaotic motions for the Duffing-like system in Eq. 3.1 are of great interest. According to the bifurcation diagrams in Figs. 3.3 and 3.3b, and the parameter map in Fig. 3.4a, for $(a, b, Q, \omega) = (1.05, 1.0, 1.1, 1.0)$ and

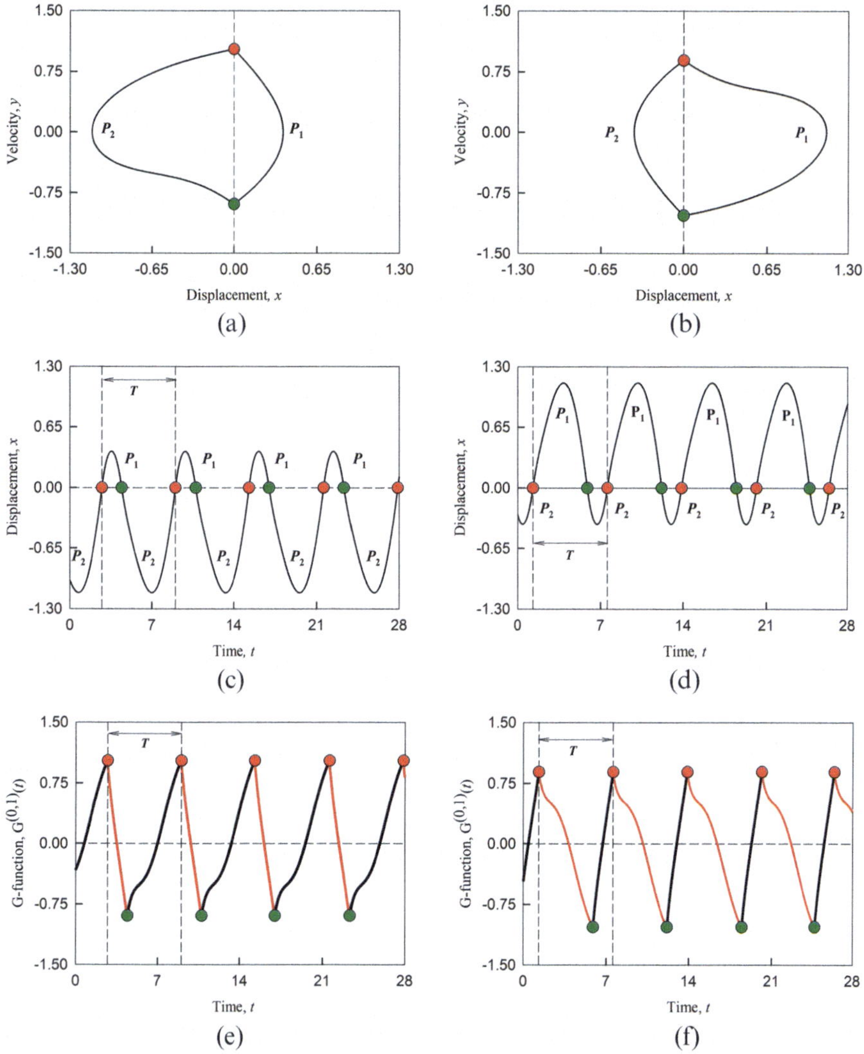

Fig. 3.5 Coexisting periodic motions with the same mappings P_{21}: (**a**) and (**b**) coexistent trajectories in phase plane; (**c**) and (**d**) displacement time history; (**e**) and (**f**) G-function time history. ($a = 1.05, b = 1, Q = 1.01, \omega = 1$; initial conditions: $t_0 = 0, x_0 = -1.0007, y_0 = -0.3248$ for (**a**), (**c**) , (**e**) and $t_0 = 0, x_0 = -0.2901, y_0 = -0.4689$ for (**b**), (**d**), (**f**))

$(t_0, x_0, y_0) = (0.0, -0.2904, -0.4681)$, the chaotic motion relative to the mapping structures of $P_{(21)^m}(m \to \infty)$ is shown in Fig. 3.6. The partial trajectory of chaotic motion in phase plane is plotted in Fig. 3.6a, and the corresponding mapping structures are also labeled. The time history responses of displacement are depicted in Fig. 3.6b, and the time histories of G-function, which examines the switchability of the Duffing-like system, are presented in Fig. 3.6c. For the red switching circles,

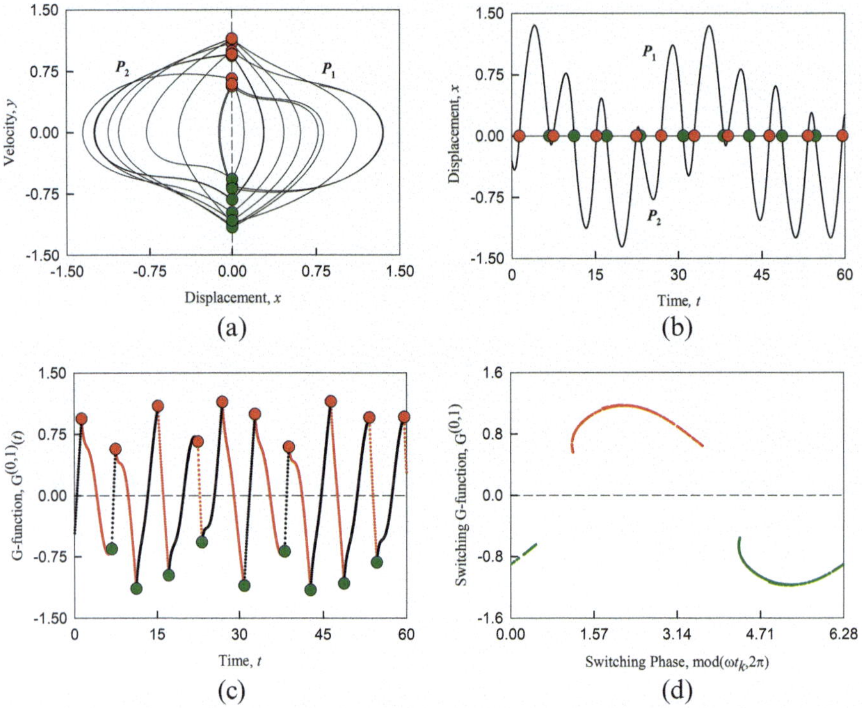

Fig. 3.6 Chaotic motion: (**a**) trajectory in phase plane; (**b**) G-function history; (**c**) G-function history; (**d**) $G^{(0,1)}$ - function versus switching phase ($a = 1.05$, $b = 1$, $Q = 1.1$, $\omega = 1$; initial condition: $t_0 = 0$, $x_0 = -0.2904$, $y_0 = -0.4681$)

the passable conditions of the system are satisfied with $G^{(0,1)}$ - function and $G^{(0,2)}$ - function positive in Eq. 3.17, which the motions at the boundary are observed from domain Ω_2 to domain Ω_1. Conversely, when $G^{(0,1)}$ - function and $G^{(0,2)}$ - function are negative, the criteria in Eq. 3.16 are also verified, which are illustrated by the green circles in Fig. 3.6c. The switching G-functions versus switching phase are depicted in switching section, as shown in Fig. 3.6d with 10000 points, which consisted of many discrete points and indicate that system lies in a chaotic state. It is also observed that the switching G-functions satisfied the passible motion conditions in Eqs. 3.16 and 3.17.

References

Min, F. H., & Chen, J. Y. (2020). The Coexisting Behaviors on the Boundary of a Duffing-like Oscillator with Signum Nonlinearity and Its FPGA-Based Implementation. *International Journal of Bifurcation and Chaos, 30*(6), 2050085.

Luo, A. C. J. (2005). A theory for non-smooth dynamic systems on the connectable domains. *Communications in Nonlinear Science and Numerical Simulation, 10*, 1–55.
Luo, A. C. J. (2008). A theory for flow switchability in discontinuous dynamical systems. *Nonlinear Analysis: Hybrid Systems, 2*(4), 1030–1061.
Luo, A. C. J. (2009). *Discontinuous Dynamical Systems on Time-varying Domains*. HEP-Springer.
Luo, A. C. J. (2012). *Discontinuous Dynamical Systems*. Higher Education Press/Springer.

Chapter 4
Switching Dynamics of a FitzHugh-Nagumo Circuit

Contents

4.1 Memristor-Based FitzHugh-Nagumo Circuit 37
4.2 Analytical Conditions for Motion Switching 40
4.3 Generic Mappings 43
4.4 Bifurcation Scenarios 45
4.5 Distribution Maps 50
4.6 Switching Dynamics 52
References 56

In this chapter, the nonautonomous FitzHugh-Nagumo neuronal circuit with piecewise memristor is investigated in flux-charge domain through the theory of switching flow. The sufficient and necessary conditions of switching motions, such as passable and grazing on the boundary, are developed to understand the switching mechanism. The switching bifurcation diagrams with changing the system parameters and initial conditions are studied to reveal the hidden extreme multistability and coexisting attractors. The parameter mappings and basins of attractor are also carried out to express the coexistences of periodic orbits and chaos with different mapping structures. The simulation results show that the switching and grazing motion are verified the effectiveness of the analytical conditions.

4.1 Memristor-Based FitzHugh-Nagumo Circuit

As in Min et al. (2021), consider a memristor-based three order FitzHugh-Nagumo neuron, as shown in Fig. 4.1a, which was proposed from the nonautonomous FitzHugh-Nagumo model. There are five discrete components including a piecewise-linear flux-controlled memristor M connected with a capacitor C, a

F. Min, *Discontinuous Dynamics and System Synchronization*, Nonlinear Systems and Complexity 39, https://doi.org/10.1007/978-3-031-66648-3_4

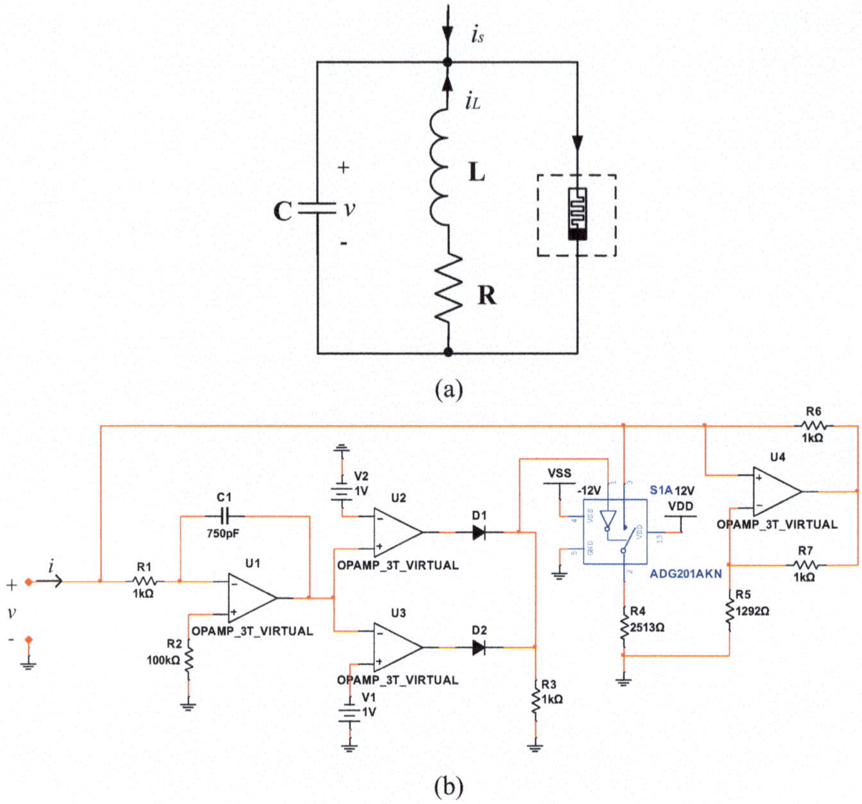

Fig. 4.1 Circuit schematic: (**a**) the memristor-based FitzHugh-Nagumo circuit; (**b**) the memristor's equivalent circuit with nonsmooth nonlinearity

inductor L, a resistor R, and an externally forced current force i_s. The memristor's equivalent circuit is shown in Fig. 4.1b, where the voltage and current of the memristor are v(t) and i(t), respectively. Although the piecewise linear model of the memristor is not realistic, the analysis can be considered, which can best understand the boundary effect of memristors.

By applying Kirchoff's laws, the circuit equations can be written as

$$\left.\begin{aligned} C\frac{dv}{dt} &= i_L - W(\varphi)v + i_s \\ L\frac{di_L}{dt} &= -Ri_L - v \\ \frac{d\varphi}{dt} &= v \end{aligned}\right\} \tag{4.1}$$

where the voltage is $v(t)$, current $i_L(t)$, flux $\varphi(t)$, time t, and i_s represents the external driving current source, $i_s = (F/\Omega)\cos(\Omega t)$, F and Ω denotes the physical amplitude and the angular frequency of the input current, respectively.

The nonsmooth memristor charge-flux (q, φ) can be described as

$$q(\varphi) = h[\varphi(t)] = b\varphi + 0.5(a - b)(|\varphi + 1| - |\varphi - 1|) \tag{4.2}$$

where φ represents the memristor flux, and the voltage-ampere characteristic of memristor is

$$\left.\begin{aligned} &i(t) = W[\varphi(t)] \cdot v(t) \\ &W(\varphi) = b + 0.5(a - b)[\text{sgn}(\varphi + 1) - \text{sgn}(\varphi - 1)] \\ &\frac{d\varphi(t)}{dt} = v(t) \end{aligned}\right\} \tag{4.3}$$

By rescaling parameters, $v = x$, $i_L = y/R$, $\varphi = z$, $t = RC\tau$, $\Omega = \omega/RC$, and redefining τ as t, the normalized Eq. 4.3 is

$$\left.\begin{aligned} &\dot{x} = y - \{\beta + 0.5(\alpha - \beta)[\text{sgn}(z + 1) - \text{sgn}(z - 1)]\} \cdot x \\ &\quad + (F_1/\omega)\cos(\omega t) \\ &\dot{y} = -\sigma x - \sigma y \\ &\dot{z} = \gamma x \end{aligned}\right\} \tag{4.4}$$

where $\alpha = aR$, $\beta = bR$, $\sigma = R^2C/L$, $\gamma = RC$ and $F_1 = R^2Cf$.

If the flow switchability theory is used to analyze the system (Eq. 4.4), the expressions of flows in different regions are the same, and the switching motion on the boundary cannot be visualized. Therefore, the system is reduced in flux-charge domain, which also helps to understand the influence of initial conditions.

From the flux-charge analysis method, the incremental flux and charge of a circuit element can be expressed as

$$\varphi(t;0) = \varphi(t) - \varphi(0) \tag{4.5}$$

$$q(t;0) = q(t) - q(0) \tag{4.6}$$

for $0 < t < \infty$. In Fig. 4.1b, the incremental variables are denoted as $\varphi_C(t;0)$, $\varphi_L(t;0)$, $\varphi_R(t;0)$, $\varphi_M(t;0)$ and $q_C(t;0)$, $q_L(t;0)$, $q_R(t;0)$, $q_M(t;0)$, which correspond to the capacitor C, inductor L, resistor R, and the memristor M. The second-order ordinary differential equations are rewritten as

$$\left.\begin{aligned} C\frac{d\varphi_C(t;0)}{dt} &= q_L(t;0) - q_M(t;0) + q_C(0) + \frac{F}{\Omega^2}\sin\Omega t \\ L\frac{dq_L(t;0)}{dt} &= -\varphi_C(t;0) - Rq_L(t;0) + \varphi_L(0) \end{aligned}\right\} \tag{4.7}$$

in which $qC(0) = Cu_C(0)$, $\varphi_L(0) = Li_L(0)$. By substituting $x(t) = \varphi_C(t;0) + \varphi_M(0)$, $y(t) = R \cdot q_L(t;0)$, $t = RC\tau$, and $\Omega = \omega/RC$ into Eq. 4.4, redefining τ as t, the normalized equations are obtained as

$$\left.\begin{aligned} \dot{x} &= y - h(x) + \frac{f}{\omega^2}\sin(\omega t) + g_1 \\ \dot{y} &= -\sigma y - \sigma x + \sigma g_2 \end{aligned}\right\} \tag{4.8}$$

where $h(x) = \beta x + 0.5(\alpha - \beta)(|x + 1| - |x - 1|)$, $\alpha = aR$, $\beta = bR$, $\sigma = R^2C/L$, $f = R^3C^2F$, $g_1 = q_M(0) + Cu_C(0) = h[\varphi_M(0)] + Cu_C(0)$, $g_2 = Li_L(0)$. The separation boundaries of this neuron model are defined as $x = 1$ and $x = -1$ due to the piecewise function $h(x)$.

4.2 Analytical Conditions for Motion Switching

From Luo (Min et al. 2021; Luo 2008, 2009, 2012), three domains of the memristive FitzHugh-Nagumo system with two boundaries are described. The passable, grazing, and sliding conditions are given, and the mappings connecting the switching sets are defined.

The system described by Eq. 4.8 is a discontinuous nonlinear dynamical system due to the piecewise function $h(x)$, in which the phase plane is divided into three connectable regions $\Omega_i (i = -1, 0, 1)$ and two separation boundaries $x = -1$ and $x = 1$ with the dashed lines labeled, as shown in Fig. 4.2. The state variables in Eq. 4.8 and the corresponding vectors for the flow are given by

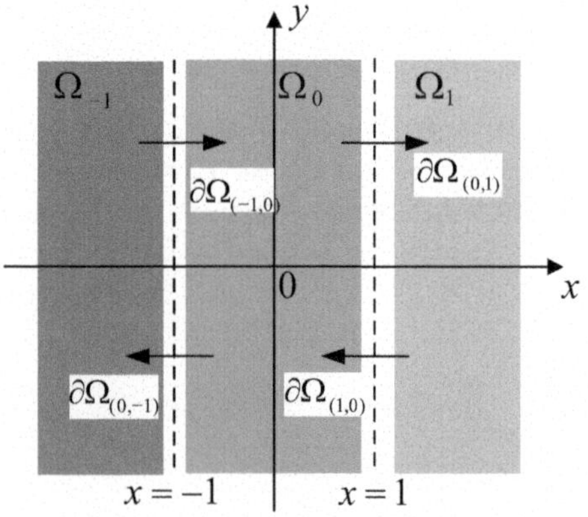

Fig. 4.2 Domain partitions in phase plane

$$\mathbf{x} = (x, y)^T$$
$$\text{and} \quad \mathbf{F} = \left(y - h(x) + \frac{f}{\omega^2} \sin(\omega t) + g_1, \; -\sigma y - \sigma x + \sigma g_2 \right)^T \tag{4.9}$$

where **F** represents the scalar force. The three named domains are defined as

$$\left.\begin{aligned} \Omega_{-1} &= \{(x, y) | x < -1\} \\ \Omega_0 &= \{(x, y) | -1 < x < 1\} \\ \Omega_1 &= \{(x, y) | x > 1\} \end{aligned}\right\} \tag{4.10}$$

and the two corresponding boundaries are

$$\left.\begin{aligned} \partial\Omega_{(0,-1)} = \partial\Omega_{(-1,0)} = \overline{\Omega}_{-1} \cap \overline{\Omega}_0 = \{(x, y) | x = -1\} \\ \partial\Omega_{(0,1)} = \partial\Omega_{(1,0)} = \overline{\Omega}_1 \cap \overline{\Omega}_0 = \{(x, y) | x = 1\} \end{aligned}\right\} \tag{4.11}$$

where Ω_i is the closure of the three domains $\Omega_i (i = -1, 0, 1)$.

In Fig. 4.2, $\partial\Omega_{(i,j)}$, $(i,j) \in \{(-1,0),(0,-1),(1,0),(0,1)\}$ means the oriented boundary between two connectable domains Ω_i and Ω_j. The system motion in each named domain considers as a continuous flow, but total forces resulting in the switching motion are discontinuous. Thus, the dynamical system in Eq. 4.9 can be rewritten as

$$\dot{\mathrm{x}} = F_{\lambda}^{(\kappa)}(\mathrm{x}, t), \; (\kappa, \lambda) \in \{(-1,0),(0,-1),(1,0),(0,1)\} \tag{4.12}$$

$$\left.\begin{aligned} &\mathbf{F}_i^{(i)}(\mathbf{x}, t) = (\dot{x}, F_i(\mathbf{x}, t))^T \text{in } \Omega_i (i \in \{-1, 0, 1\}) \\ &\mathbf{F}_i^{(j)}(\mathbf{x}, t) = (\dot{x}, F_j(\mathbf{x}, t))^T \\ &\qquad \textit{in } \Omega_i ((i,j) \in \{(-1,0),(0,-1),(1,0),(0,1)\}) \\ &\mathbf{F}_1^{(1)}(\mathbf{x}, t) = (1, y)^T \textit{or } \mathbf{F}_{-1}^{(-1)}(\mathbf{x}, t) = (-1, y)^T \\ &\qquad \textit{on } \partial\Omega_{(i,j)} \text{for stick}, \\ &\mathbf{F}_1^{(1)}(\mathbf{x}, t), \mathbf{F}_{-1}^{(-1)}(\mathbf{x}, t) \in \left[\mathbf{F}_i^{(i)}(\mathbf{x}, t), \mathbf{F}_j^{(j)}(\mathbf{x}, t)\right] \\ &\qquad \textit{on } \partial\Omega_{(i,j)} \text{for non-stick} \end{aligned}\right\} \tag{4.13}$$

where

$$\mathbf{F}_i(\mathbf{x}, t) = y - \left[\beta^{(i)} x + 0.5\left(\alpha^{(i)} - \beta^{(i)}\right)(|x+1| - |x-1|\right] + (f/\omega^2)\sin(\omega t) \tag{4.14}$$

with $\kappa \neq 0$ and $\lambda \neq 0$ in Eq. 4.14—the subscript λ and superscript κ stand for two adjacent regions for $(i,j) \in \{(-1,0),(0,-1),(1,0),(0,1)\}$. $\mathbf{F}_i^{(i)}(\mathbf{x}, t)$ is the real vector field in i-domain, and $\mathbf{F}_i^{(j)}(\mathbf{x}, t)$, controlled by the vector field in j-domain, is the fictitious vector field in i-domain. $\mathbf{F}_1^{(1)}(\mathbf{x}, t)$ and $\mathbf{F}_{-1}^{(-1)}(\mathbf{x}, t)$ are the vector fields on two separation boundaries. $\mathbf{F}_i(\mathbf{x}, t)$ is the scalar force in i-domain, which is given by Eq. 4.14. Then the discontinue behaviors for the whole memristive FitzHugh-Nagumo system in Eq. 4.8 can be developed by the expression in Eq. 4.14.

From the theory of flow switchability in Luo (Min et al. 2021; Luo 2008, 2009), the existence conditions for passable motion on the boundary $\partial\Omega_{ij}$ at time t_m are given by

$$\left.\begin{aligned} n_{\partial\Omega_{(i,j)}}^T \cdot \mathbf{F}^{(i)}\left(\mathbf{x}_m^{(i)}, t_{m\pm}\right) < 0 \\ n_{\partial\Omega_{(i,j)}}^T \cdot \mathbf{F}^{(j)}\left(\mathbf{x}_m^{(j)}, t_{m\pm}\right) < 0 \end{aligned}\right\} \text{for } (i,j) \in \{(1,0), (0,-1)\} \tag{4.15}$$

$$\left.\begin{aligned} n_{\partial\Omega_{(i,j)}}^T \cdot \mathbf{F}^{(i)}\left(\mathbf{x}_m^{(i)}, t_{m\pm}\right) > 0 \\ n_{\partial\Omega_{(i,j)}}^T \cdot \mathbf{F}^{(j)}\left(\mathbf{x}_m^{(j)}, t_{m\pm}\right) > 0 \end{aligned}\right\} \text{for } (i,j) \in \{(-1,0), (1,0)\} \tag{4.16}$$

where $\mathbf{F}^{(i)}(\mathbf{x}, t)$ represents the real vector field in i-domain and the imaginary in j-domain. The switching time t_m denotes the system motion on the boundary, and t_{m+} and t_{m-} are for the motion infinite approaching and just leaving the boundary, respectively.

The conditions for grazing response on the boundary at time t_m are

$$\left.\begin{aligned} n_{\partial\Omega_{(i,j)}}^T \cdot \mathbf{F}^{(i)}\left(\mathbf{x}_m^{(i)}, t_{m\pm}\right) = 0 \\ n_{\partial\Omega_{(i,j)}}^T \cdot D\mathbf{F}^{(i)}\left(\mathbf{x}_m^{(i)}, t_{m\pm}\right) > 0 \end{aligned}\right\} \text{for } (i,j) \in \{(1,0), (0,-1)\} \tag{4.17}$$

$$\left.\begin{aligned} n_{\partial\Omega_{(i,j)}}^T \cdot \mathbf{F}^{(i)}\left(\mathbf{x}_m^{(i)}, t_{m\pm}\right) = 0 \\ n_{\partial\Omega_{(i,j)}}^T \cdot D\mathbf{F}^{(i)}\left(\mathbf{x}_m^{(i)}, t_{m\pm}\right) < 0 \end{aligned}\right\} \text{for } (i,j) \in \{(-1,0), (1,0)\} \tag{4.18}$$

in domain Ω_i, where $D\mathbf{F}^{(i)}(\mathbf{x}, t) = (F_i(\mathbf{x}, t), \nabla F^{(i)}(\mathbf{x}, t) \cdot \mathbf{F}^{(i)}(\mathbf{x}, t) + \partial_t F_\alpha(\mathbf{x}, t))^T$.

The conditions for sliding response on the boundary at time t_m are

$$\left.\begin{aligned} \mathbf{n}_{\partial\Omega_{(i,j)}}^T \cdot \mathbf{F}^{(i)}(\mathbf{x}_m, t_{m-}) < 0 \\ \mathbf{n}_{\partial\Omega_{(i,j)}}^T \cdot \mathbf{F}^{(j)}(\mathbf{x}_m, t_{m+}) > 0 \end{aligned}\right\} \text{for } n_{\partial\Omega_{(i,j)}} \to \Omega_i \tag{4.19}$$

where $(i,j) \in \{(-1,0), (0,-1), (1,0), (0,1)\}$. There is no sliding response on two separation boundaries for this memristive FitzHugh-Nagumo system in Eq. 4.8.

As two oriented boundaries depend on $\varphi(x, y) = x \pm 1 = 0$, the normal vectors of the two boundaries are

$$n_{\partial\Omega_{(-1,0)}} = n_{\partial\Omega_{(1,0)}} = (1,0)^{\mathrm{T}} \tag{4.20}$$

According to the above equations (Eqs. 4.15–4.20), the zero and first G-function are

$$\left.\begin{aligned} G^{(0,j^{(k)})}(\mathbf{x}_m, t_{m\pm}) &= \mathbf{n}^T_{\partial\Omega_{(i,j)}} \cdot \mathbf{F}^{(i)}(\mathbf{x}_m, t_{m\pm}) = \mathbf{F}_i(\mathbf{x}_m, t_{m\pm}), \\ G^{(1,j^{(k)})}(\mathbf{x}_m, t_{m\pm}) &= \mathbf{n}^T_{\partial\Omega_{(i,j)}} \cdot D\mathbf{F}^{(i)}(\mathbf{x}_m, t_{m\pm}) = D\mathbf{F}_i(\mathbf{x}_m, t_{m\pm}) \\ &= \nabla F_i(\mathrm{x}, t) \cdot \mathbf{F}^{(i)}(\mathbf{x}, t) + \partial_t F_i(\mathbf{x}, t)|_{(\mathbf{x}_m, t_{m\pm})} \end{aligned}\right\} \tag{4.21}$$

with $(j, k) \in \{1, 2\}$.

4.3 Generic Mappings

To make a better understanding of the switching motions of system in Eq. 4.8, the switching sets and mappings of the system are illustrated in Fig. 4.3, where the switching sets are defined as the set containing the switching points and switching time and can be obtained from the switching boundaries. For two boundaries $x = -1$ and $x = 1$, the switching sets, represented by the dashed lines, are given as

$$\left.\begin{aligned} \Xi_1^+ &= \{(x_k, y_k, t_k) | x_k = 1^+\} \\ \Xi_1^- &= \{(x_k, y_k, t_k) | x_k = 1^-\} \\ \Xi_{-1}^+ &= \{(x_k, y_k, t_k) | x_k = -1^+\} \\ \Xi_{-1}^- &= \{(x_k, y_k, t_k) | x_k = -1^-\} \end{aligned}\right\} \tag{4.22}$$

The superscript "+" or "−" implies that the switching set is infinitely close to the right or left side of the separation boundary. Thus, six mappings are given by

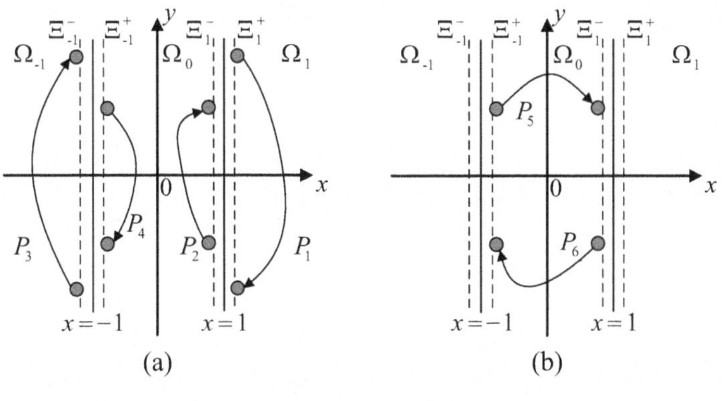

Fig. 4.3 Switching sets and mappings: (**a**) local mappings; (**b**) global mappings

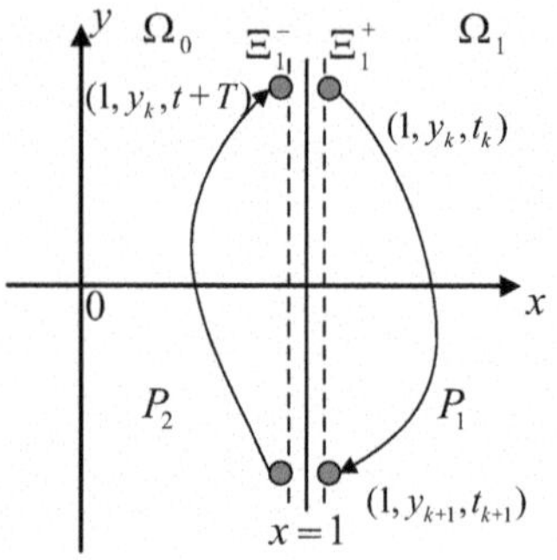

Fig. 4.4 A periodic response of P_{12}

$$\left.\begin{array}{ll} P_1 : \Xi_1^+ \to \Xi_1^+ & P_2 : \Xi_1^- \to \Xi_1^- \\ P_3 : \Xi_{-1}^- \to \Xi_{-1}^- & P_4 : \Xi_{-1}^+ \to \Xi_{-1}^+ \end{array}\right\} \text{for local mappings} \tag{4.23}$$

$$P_5 : \Xi_1^- \to \Xi_{-1}^+ \qquad P_6 : \Xi_{-1}^+ \to \Xi_1^- \text{ for global mappings} \tag{4.24}$$

The local mappings, $P_l(l = 1, 2, 3, 4)$ in Fig. 4.3a, map the initial condition $(\pm 1, y_k, t_k)$ to the final state $(\pm 1, y_{k+1}, t_{k+1})$ in the same switching set, and the global mappings, $P_l(l = 5, 6)$ in Fig. 4.3b, denote the initial condition $(\mp 1, y_k, t_k)$ from one switching set to the other to arrive the final state $(\pm 1, y_{k+1}, t_{k+1})$.

The algebraic equations of four local mappings $P_l(l = 1, 2, 3, 4)$ are expressed as

$$\left.\begin{array}{l} f_1^{(l)} \equiv \pm 1 - \Phi_1^{(i)}(\pm 1, y_k, t_{k+1}, t_k) = 0 \\ f_2^{(l)} \equiv y_{k+1} - \Phi_2^{(i)}(\pm 1, y_k, t_{k+1}, t_k) = 0 \end{array}\right\} \tag{4.25}$$

and for $P_l(l = 5, 6)$,

$$\left.\begin{array}{l} f_1^{(l)} \equiv \pm 1 - \Phi_1^{(i)}(\mp 1, y_k, t_{k+1}, t_k) = 0 \\ f_2^{(l)} \equiv y_{k+1} - \Phi_2^{(i)}(\mp 1, y_k, t_{k+1}, t_k) = 0 \end{array}\right\} \tag{4.26}$$

where $\Phi^{(i)} = \left(\Phi_1^{(i)}, \Phi_2^{(i)}\right)^T$.

As an example of the periodic flow of $P_{21} = P_1 \circ P_2$, its periodic orbit is illustrated in Fig. 4.4. The periodic behavior in one period can be illustrated by mapping structures as follows: first, at the starting point $(1, y_k, t_k) \in \Xi_1^+$, the flow in domain Ω_1 gets to the point $(1, y_{k+1}, t_{k+1}) \in \Xi_1^+$ in the same switching set through mapping P_1 at time t_{k+1}. Then, as the passable condition of the flow in Eq. 4.20 being satisfied, the point will switch across the boundary $x = 1$ in domain Ω_0. Afterward, the final state $(1, y_{k+1}, t_{k+1}) \in \Xi_1^-$ turns into the initial condition for mapping P_2 at the switching set Ξ_1^-.

Therefore, during N-periods of excitation, if a periodic motion of $\mathbf{x}_{k+2} = \mathbf{P}x_k$ in $\mathbf{x}_k = (y_k, \omega t_k)^T$ happens, the periodicity of the periodic motion will be

$$x_{k+2} = x_k, t_{k+2} = t_k + T \tag{4.27}$$

where T is the period for this periodic flow. The trajectory of the periodic motion for other periodic and chaotic responses can be analyzed similarly.

4.4 Bifurcation Scenarios

In this section, the bifurcation diagram and Lyapunov exponents will be plotted to analyze the boundary dynamics characteristics of the FitzHugh-Nagumo system.

Note that the acronyms "GB," "PD," "SN," and "PF" labeled in the bifurcation diagram stand for "grazing bifurcation," "period-doubling bifurcation," "saddle-node bifurcation," and "pitchfork bifurcation," and the mapping structures of the periodic responses are also marked.

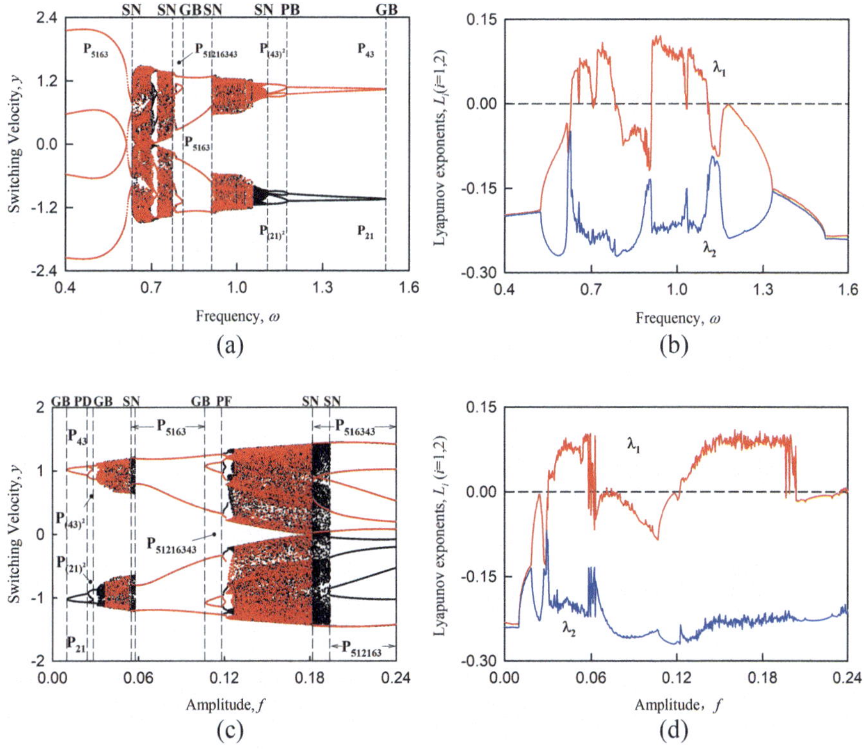

Fig. 4.5 Bifurcation and Lyapunov spectrum with $\omega \in [0.4, 1.6]$**:** (**a**) Bifurcation diagram; (**b**) Lyapunov exponent spectrum, with $f \in [0.00, 0.24]$; (**c**) bifurcation diagram; (**d**) Lyapunov exponent spectrum

Table 4.1 Coexisting mapping structures of the parameter ω

Ranges of ω	Coexisting mapping structures
[1.052, 1.108]	$P_{(21)^{\infty}}$ with $P_{(43)^{\infty}}$
[1.108, 1.174]	$P_{(21)^2}$ with $P_{(43)^2}$
[1.174, 1.519]	P_{21} with P_{43}
Ranges of f	**Coexisting mapping structures**
[0.0102, 0.0240]	P_{21} with P_{43}
[0.0240, 0.0283]	$P_{(21)^2}$ with $P_{(43)^2}$
[0.182, 0.194]	$P_{(51216343)^{\infty}}$ with P_{516343}
[0.194, 0.24]	P_{512163} with P_{516343}

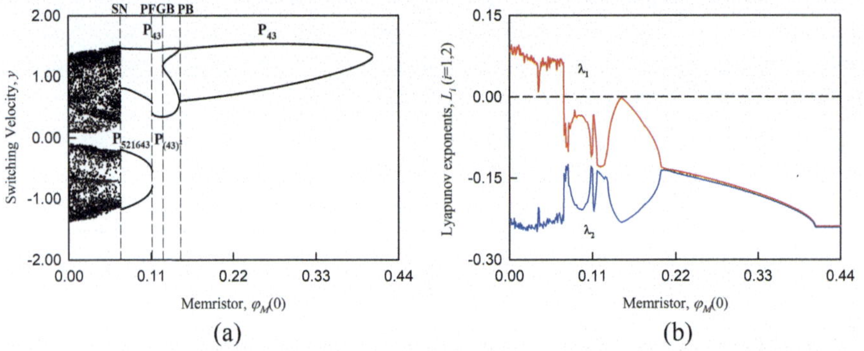

Fig. 4.6 Bifurcation and Lyapunov spectrum with $\varphi_M(0) \in [0.0, 0.44]$: (**a**) bifurcation diagram; (**b**) Lyapunov exponent spectrum

First, set the initial conditions $u_C(0) = 0.00$, $\varphi_M(0) = 0.00$, and $i_L(0) = 0.00$ and the parameters $\alpha = -1.03$, $\beta = -0.5$, $\sigma = 0.98$, $f = 0.15$. The bifurcation diagram of frequency ω is presented in Figs. 4.5a, and 4.5b shows its corresponding Lyapunov exponents. As the frequency ω increases, the system tends to be in period eventually. The FitzHugh-Nagumo system goes from period to chaos by saddle-node bifurcation at $\omega = 0.633$ and then turns into period at $\omega = 0.773$. The grazing bifurcation happens at $\omega = 0.810$, the mapping structure of which is P_{5163}. The neuron model turns back into chaos by saddle-node bifurcation at $\omega = 0.910$ and finally turns to be period by a series of bifurcation, such as saddle-node bifurcation at $\omega = 1.108$, period-doubling bifurcation at $\omega = 1.174$, and grazing bifurcation at $\omega = 1.519$. There is blank region in [1.519, 1.6], which means that the relevant phase plane doesn't switch across the two separation boundaries $x = 1$ and $x = -1$. Moreover, the bifurcation and its Lyapunov exponents diagram for amplitude f are depicted in Fig. 4.5c, d. The bifurcation graph starts from the periodic motion of rubbing the edges and passes through period-doubling bifurcation, saddle-node bifurcation to chaos, and after experiencing period and chaos, it finally tends to period. It can also be observed that chaos only occurs in some specific regions. The

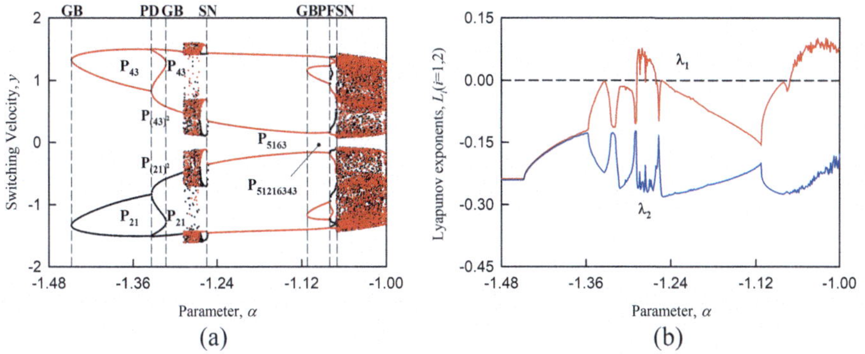

Fig. 4.7 Bifurcation and Lyapunov spectrum with $\alpha \in [-1.48, -1.0]$: (**a**) bifurcation diagram; (**b**) Lyapunov exponent spectrum

coexisting behaviors of symmetric attractors and the ranges of frequency ω and amplitude f are also listed in Table 4.1.

Next, by varying $\varphi_M(0) \in [0.00, 0.44]$, the bifurcation diagram and Lyapunov exponent spectrum can be simulated from Eq. 4.10 as shown in Fig. 4.6, where $\alpha = -1.03$, $\beta = -0.5$, $\sigma = 0.98$, $f = 0.15$, $\omega = 0.75$, and the initial conditions are $u_C(0) = 0.00$ and $i_L(0) = 0.00$. From Fig. 4.6a, the memristive system is in chaos when $\varphi_M(0) \in (0.0, 0.069)$ with $P_{(51216343)^\infty}$, which means the trajectories in the chaotic phase plane switch through both the boundaries $x = 1$ and $x = -1$. And then the system gets into period-1 with P_{5163} via saddle-node bifurcation at $\varphi_M(0) = 0.069$, containing the global mapping structures P_5 and P_6. At $\varphi_M(0) = 0.111$, the pitchfork bifurcation happens, resulting in the mapping structure P_{43}. The grazing bifurcation occurs at $\varphi_M(0) = 0.126$ and the system is in $P_{(43)^2}$. The system in Fig. 4.1 turns back into the mapping structures P_{43} through reverse period-doubling bifurcation at $\varphi_M(0) = 0.1495$. The bifurcation route disappears at $\varphi_M(0) = 0.406$, which means the corresponding phase plane does not switch the two separation boundaries. The Lyapunov exponent, in Fig. 4.6b, is consistent with the bifurcation diagram.

And then, the influence of α on the system dynamic behaviors can be observed in the range of $\alpha \in [-1.48, -1.0]$, which the initial conditions of the capacitor C, inductor L, and memristor M of Eq. 4.8 are set to zero. The bifurcation and Lyapunov exponents are depicted in Fig. 4.7, respectively, where $\beta = -0.5$, $\sigma = 0.98$, $f = 0.15$, $\omega = 0.75$, the initial condition $(t_0, x_0, y_0) = (0.0, 0.001, 0.0)$, and the orbits in blue stands for α increasing and red for decreasing. Observing the bifurcation diagram of parameter α in Fig. 4.7a, the blank region for $\alpha \in (-1.500, -1.448)$ means that the trajectories in phase planes switch neither of the two boundaries ($x = 1$, $x = -1$). Because the blue and red orbits for the changing parameter are similar and complementary, the bifurcation route for the increasing α is depicted in detail. The grazing

Table 4.2 Coexisting mapping structures of the parameter α

Ranges of α	Coexisting mapping structures	Ranges of α	Coexisting mapping structures
$(-1.448, -1.334)$	P_{21} with P_{43}	$(-1.313, -1.288)$	P_{21} with P_{43}
$(-1.334, -1.313)$	$P_{(21)^2}$ with $P_{(43)^2}$	$(-1.080, -1.070)$	$P_{51216343}$ with $P_{51216343}$

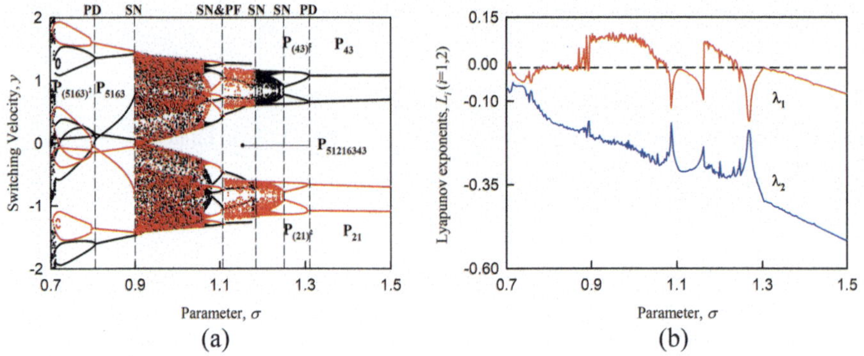

Fig. 4.8 Bifurcation and Lyapunov spectrum with $\sigma \in [0.7, 1.5]$: (**a**) Bifurcation diagram; (**b**) Lyapunov exponent spectrum

bifurcation happens at $\alpha = -1.448$, the trajectory of which in phase plane is tangent to the boundary. Then the system is in period-1 with the mapping structure P_{21} until the period-doubling bifurcation occurs at $\alpha = -1.334$, which leads to the period-2 with the mapping structure $P_{(21)^2}$. Because of the grazing bifurcation at $\alpha = -1.313$, the motion of the system turns back into period-1 with P_{21}. For $\alpha \in [-1.288, -1.255]$, the motion is in chaos with $P_{(5163)^\infty}$, which switch through both boundaries $x = -1$ and $x = 1$. Via the saddle-node bifurcation at $\alpha = -1.255$, the neuronal system changes into period-1 with P_{5163}, which is global periodic response. Finally, the motion of system (Eq. 4.8) goes into chaotic state through a series of bifurcations, like the grazing bifurcation at $\alpha = -1.112$, the pitchfork bifurcation at $\alpha = -1.080$, and the saddle-node bifurcation at $\alpha = -1.070$. The coexisting behaviors of the memristor-based nonsmooth FitzHugh-Nagumo system and the ranges of the parameter α are listed in Table 4.2.

Lastly, fix $\alpha = -1.03, \beta = -0.5, f = 0.15, \omega = 0.75$ with varying parameter σ—the bifurcation diagram for $\sigma \in [0.7, 1.5]$ and the corresponding Lyapunov exponent are observed in Fig. 4.8, where the initial conditions $(t_0, x_0, y_0) = (0, 001, 0.0)$ for both increasing and decreasing σ. From Fig. 4.8a, there are various coexisting attractors with different mapping structures. For $\sigma \in [0.700, 1.065]$, two bifurcation routes change from period to chaos via period-doubling bifurcation at $\sigma = 0.805$ and saddle-node bifurcation at $\sigma = 0.899$. Note that for $\sigma \in [1.107, 1.185]$, periodic and chaotic response coexist with $P_{51216343}$ and $P_{(51216343)^\infty}$, and then two coexisting

Table 4.3 Coexisting mapping structures of the parameter σ

Ranges of σ	Coexisting mapping structures	Ranges of σ	Coexisting mapping structures
(0.700, 0.805)	$P_{(5163)^2}$ with $P_{(5163)^2}$	(1.185, 1.251)	$P_{(43)^\infty}$ with $P_{(21)^\infty}$
(0.805, 0.809)	P_{5163} with P_{5163}	(1.185, 1.121)	$P_{(43)^2}$ with $P_{(21)^2}$
(1.107, 1.185)	$P_{51216343}$ with $P_{51216343}$	(1.121, 1.500)	P_{43} with P_{21}

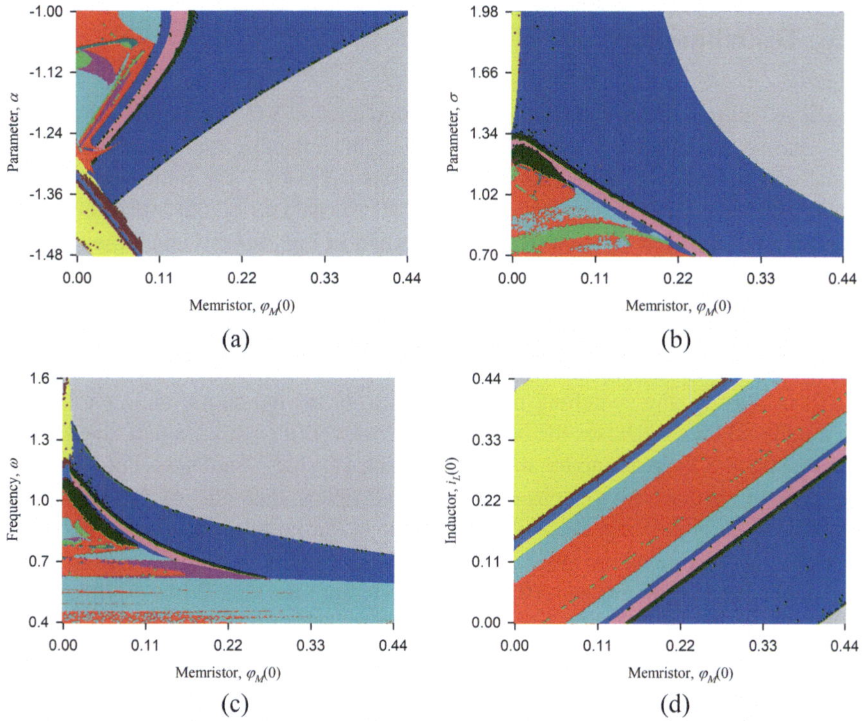

Fig. 4.9 The distribution maps of different motions: (**a**) on $\varphi_M(0) - \alpha$; (**b**) on $\varphi_M(0) - \sigma$; (**c**) on $\varphi_M(0) - f$; (**d**) on $\varphi_M(0) - i_L(0)$

chaotic attractors with $P_{(43)^\infty}$ and $P_{(21)^\infty}$ appear in $\sigma \in [1.185, 1.121]$. After that, the neuronal model turns back into period, and its corresponding mapping structures for the two kinds of the coupled periodic coexisting attractors are $P_{(43)^2}$ with $P_{(21)^2}$ and P_{43} with P_{21}. The Lyapunov exponent in Fig. 4.8b is in consistent with the bifurcation of increasing σ. In Table 4.3 is the coexisting ranges of the parameter σ and the corresponding mapping structures of the system Eq. (4.8).

Table 4.4 Colorful ribbons and its corresponding mapping structures

Color	Mapping structure	Color	Mapping structure
	P_{21}		P_{43}
	P_{5163}		$P_{(43)^2}$
	P_{516343}		$P_{(21)^\infty}$
	$P_{(21)^2}$		$P_{(43)^\infty}$
	P_{512163}		$P_{(5163)^\infty}$ or $P_{(51216343)^\infty}$
	$P_{(5163)^2}$ or $P_{51216343}$		No switching motion

4.5 Distribution Maps

The distribution maps of different motions, simulated by VC^{++}, are drawn to display the periodic and chaotic motions in a global sight, reflecting the sensitivity on the initial condition $\varphi_M(0)$. According to conditions 4.15–4.19, the numerical simulations, with the computational error tolerance 10^{-12}, are carried out under the initial condition $(t_0, x_0, y_0) = (0.0, 0.001, 0.0)$, as shown in Fig. 4.9, in which the discrete points stand for the coexisting different behaviors on initial conditions. The colorful ribbons and the mapping structures are listed in Table 4.4.

In Fig. 4.9a, b, the distribution maps of different motions are simulated on the planes of memristor and parameter α and σ. With the initial value of the memristor $\varphi_M(0)$ increasing, the switching motion tends to be periodic and monotonous and eventually no longer crosses the boundaries. However, if $\varphi_M(0)$ is small, the motion types in Eq. 4.8 keep abundant and variable with varying parameters. For example, the chaotic motion can be observed, which implies that one of the important conditions for generating chaos in this circuit is the initial condition being small enough. For big $\varphi_M(0)$, the system keeps in P_{43} or even no switching motion.

The distribution maps on $\varphi_M(0) - \omega$ are plotted in Fig. 4.9c. When $\varphi_M(0)$ goes to a certain value, as the frequency ω increases, the FitzHugh-Nagumo system will eventually become a periodic motion without crossing the boundaries ($x = 1$, $x = -1$), which is also consistent with the frequency characteristics of the system's external excitation. Only when the system frequencies ω and $\varphi_M(0)$ are both small enough that the system will show rich boundary dynamic behavior.

The distribution maps of different motions on the planes of the memristor and inductor $\varphi_M(0) - i_L(0)$ are illustrated in Fig. 4.9d. For $\varphi_M(0)$ increasing, the system changes from chaotic to periodic and the corresponding phase planes move eventually to the right boundary $x = 1$ when the inductor $i_L(0)$ is little enough. Hence, to guarantee abundant dynamics, the initial conditions should be minimal.

As an example, the coexisting attractors in phase planes, by varying $\varphi_M(0)$, are depicted in Fig. 4.10a, b, where $\alpha = -1.03$, $\beta = -0.5$, $\sigma = 0.98$, $f = 0.15$, $\omega = 0.75$, and $u_C(0) = 0.0$, $i_L(0) = 0.0$. For $\varphi_M(0)$ increasing, crossing two boundaries from right to left, the system's motion state changes from chaotic to periodic response, and it gradually crosses the boundary $x = -1$—finally it does not switch across the

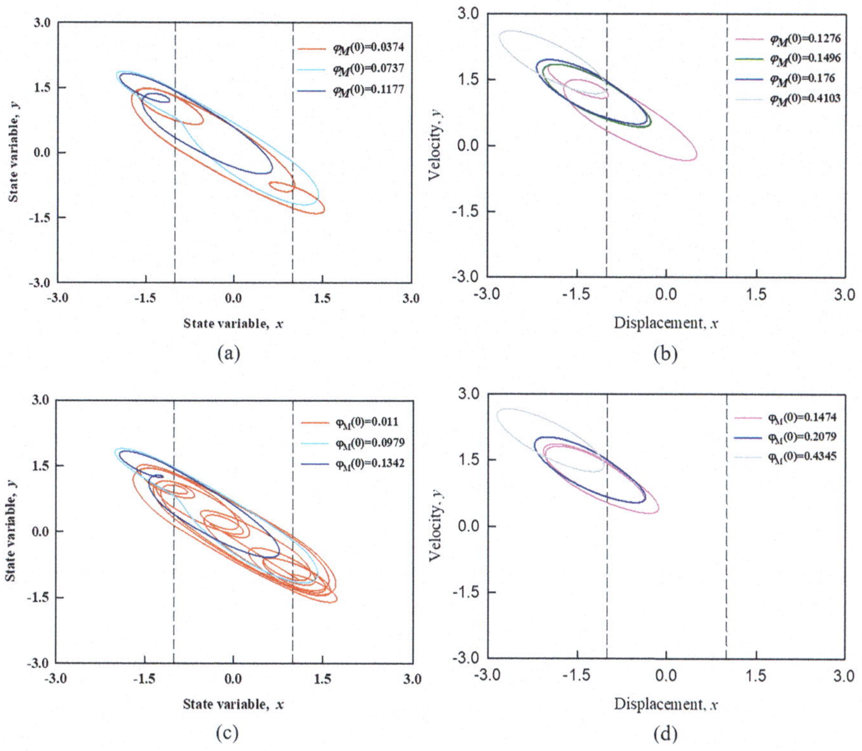

Fig. 4.10 Phase planes of coexisting attractors with varying $\varphi_M(0)$: (**a**) and (**b**) $\alpha = -1.03$; (**c**) and (**d**) $i_L(0) = 0.022$

separation boundaries any more. These phase planes are consistent with the corresponding distribution maps on $\varphi_M(0) - \alpha$. Now, the initial conditions are changed to $u_C(0) = 0.0$, $i_L(0) = 0.022$—the coexistence of period and chaotic attractors is simulated as shown in Fig. 4.10c, d, where the switching motion changes from switching two boundaries to one boundary and finally does not pass through any boundary. These switching motion behaviors are consistent with the distribution maps on $\varphi_M(0) - i_L(0)$.

Moreover, the nonsmooth memristive FitzHugh-Nagumo circuit in Fig. 4.1 is also sensitive to parameter values, and the parameter maps show how the mapping structures of periodic and chaotic responses change with varying the parameter values. Diverse colors represent different mapping structures listed in Table 4.4, and the initial conditions are set to $(t_0, x_0, y_0) = (0.0, 0.001, 0.0)$ in Fig. 4.11. As can be seen from Fig. 4.11, with $\alpha - \beta$ plane, for $\alpha < -1.45$, the motion doesn't switch through boundaries in Fig. 4.11a. For $\alpha > -1.45$, with β increasing, different mapping structures can be observed, and the chaotic region is stable for $\alpha > -1.09$ and $\beta > -0.65$. In Fig. 4.11b with $\sigma - f$ plane, for big parameters, the system presents the periodic state, and for $\sigma < 1.34$, the system possesses rich dynamical behaviors, including chaotic responses $P_{(21)^\infty}$, $P_{(43)^\infty}$, and $P_{(21)^\infty(43)^\infty}$.

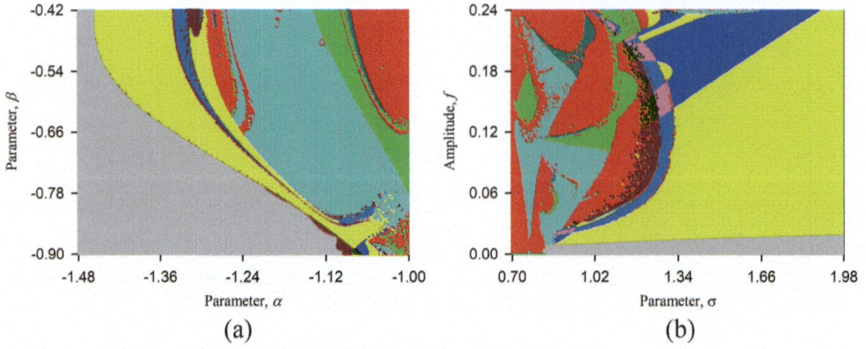

Fig. 4.11 Parameter maps: (**a**) $\alpha - \beta(\sigma = 0.98, f = 0.15, \omega = 0.75)$; (**b**) $\sigma - f(\alpha = -1.03, \beta = -0.5, \sigma = 0.98)$

4.6 Switching Dynamics

The periodic and chaotic motion of the neural FitzHugh-Nagumo circuit are presented for an understanding of motion switchabilities at two separation boundaries $x = 1$ and $x = -1$. Equation 4.8, at the boundary $x = 1$, the switching points for G-function $G^{(0,1)}$ and $G^{(0,2)}$ are equal—at $x = -1$, the switching points for G-function $G^{(0,3)}$ and $G^{(0,4)}$ are also equal. The boundary analysis for some coexisting attractors is given in Figs. 4.12, 4.13 and 4.14, where the white symbols stand for Ω_0 switching to Ω_1; green, Ω_1 switching to Ω_0; orange, Ω_{-1} switching to Ω_0; coffee, Ω_0 switching to Ω_{-1}; and gray, the starting point for periodic response. According to the bifurcation diagrams in Fig. 4.8a, the periodic response and chaotic response coexist under different initial conditions when $\alpha = -1.03$, $\beta = -0.5$, $\sigma = 1.12$, $f = 0.15$, $\omega = 0.75$. Set the initial conditions of $t_0 = 0.0$, $x_0 = 1.006$, $y_0 = -1.235$ and $t_0 = 0.0$, $x_0 = -2.333$, $y_0 = 2.210$—the coexistence of mapping structure $P_{51216343}$ and $P_{(123456)^\infty}$ is simulated in Figs. 4.13 and 4.14.

First, the trajectory of periodic orbit with the mapping $P_{51216343}$ is depicted in Fig. 4.12a, which has no stick motion on the boundaries. The time history of state variable x is presented in Fig. 4.12b, and the time histories of G-functions are, respectively, shown in Fig. 4.12c, d, where one period T of motion is marked. The G-function is the real flow in the corresponding domain. Exampled that in domain Ω_1, $G^{(0,1)}$-function is real flow and $G^{(0,2)}$-function is imaginary flow. The real and imaginary flows in other domains can be identified similarly. Observing Fig. 4.12a–d in one period T, from the starting point at the boundary $x = -1$, the trajectory in the phase plane first gets to the switching set Ξ_1^- about the boundary $x = 1$ by global mapping P_5 and then switched through the boundary to Ξ_1^+, from domain Ω_0 to Ω_1, satisfied with $G^{(0,1)} = G^{(0,2)} > 0$. The trajectory goes through the boundary $x = 1$ from domain Ω_1 to Ω_0, satisfied with $G^{(0,1)} = G^{(0,2)} < 0$. After another switching motion about the boundary $x = 1$, the trajectory then moves to Ξ_{-1}^+ approaching the boundary $x = -1$ by global mapping P_6. Afterward, the trajectory switches through

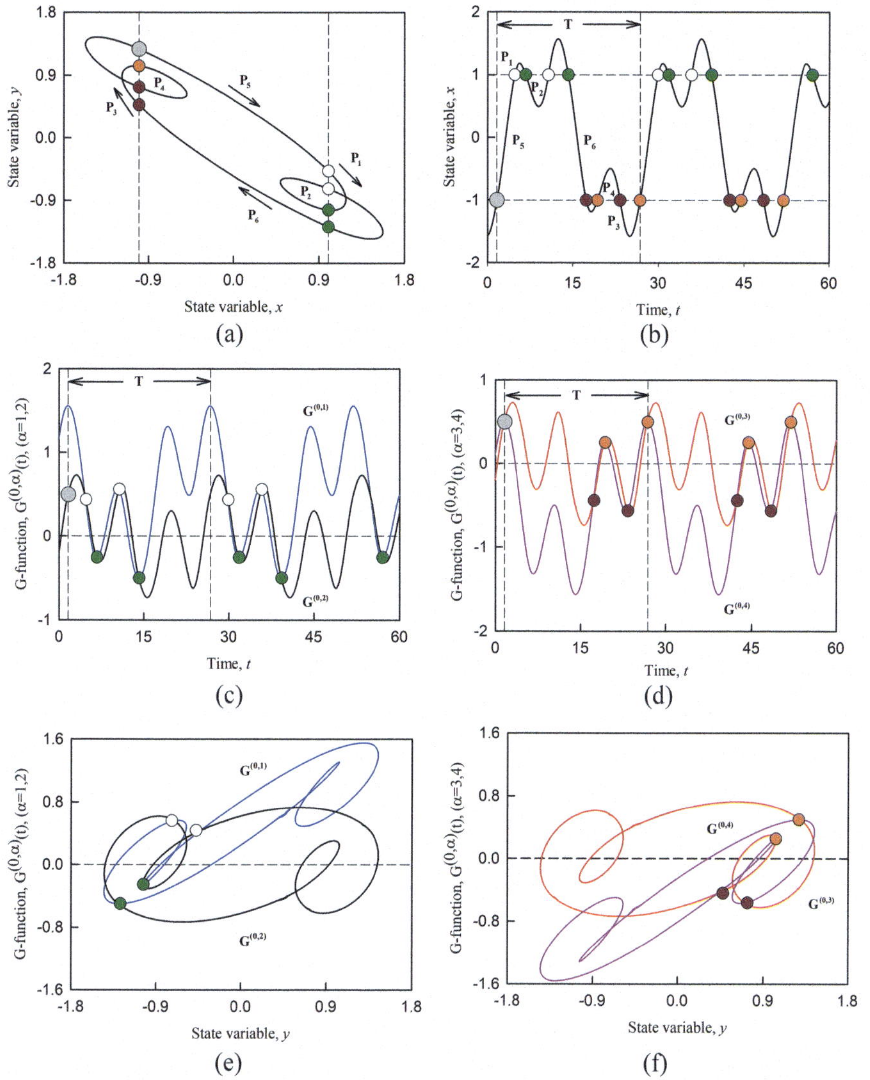

Fig. 4.12 Periodic motion of $P_{51216343}$: (**a**) phase plane; (**b**) time history of state variable x; (**c**) and (**d**) time histories of $G^{(0,1)}$-,$G^{(0,2)}$-,$G^{(0,3)}$-, and $G^{(0,4)}$-functions; (**e**) and (**f**) $G^{(0,1)}$-, $G^{(0,2)}$-, $G^{(0,3)}$-, and $G^{(0,4)}$-function distribution along state variable y

the boundary $x = -1$ to Ξ_{-1}^{-}, from domain Ω_0 to Ω_{-1}, satisfied with $G^{(0,3)} = G^{(0,4)} < 0$, and then goes through the boundary $x = -1$ from domain Ω_{-1} to Ω_0, satisfied with $G^{(0,3)} = G^{(0,4)} > 0$. After another switching motion about the boundary $x = -1$, the trajectory goes back to the starting point. The depicted behaviors of the periodic motion are in agreement with Eqs. 4.15 and 4.16. The four

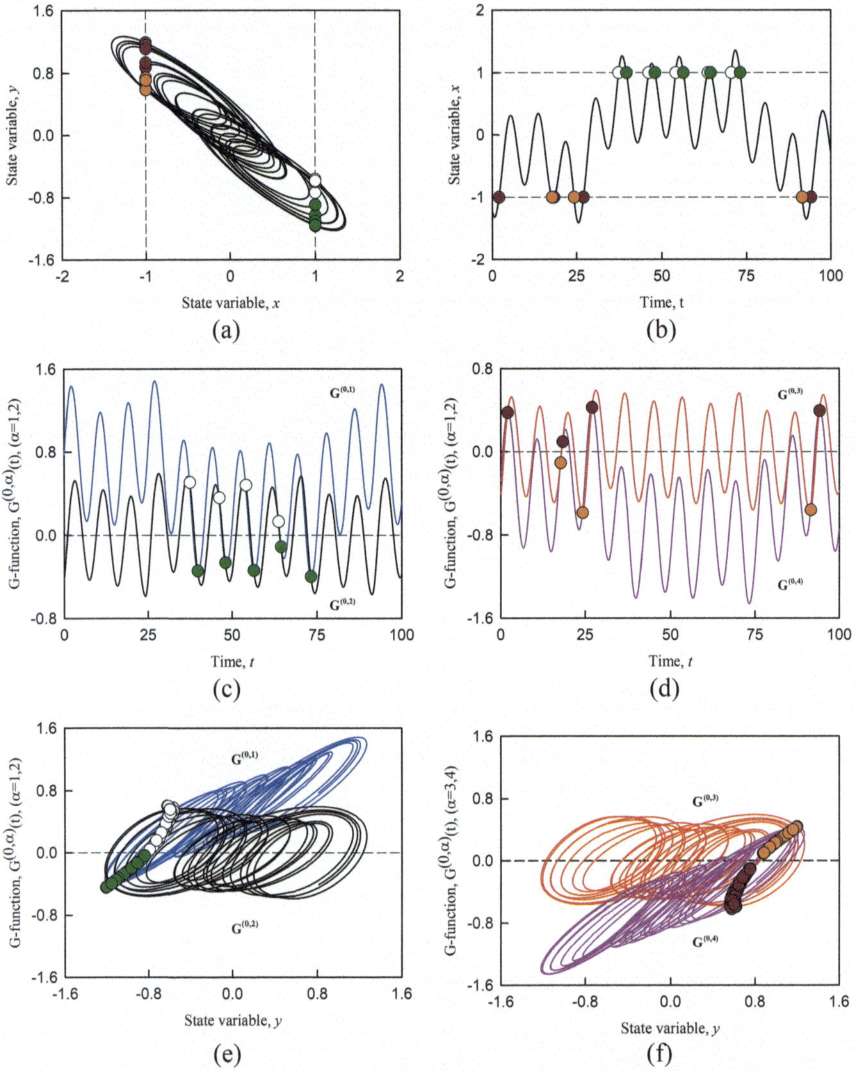

Fig. 4.13 Chaotic motion of $P_{(21)^{\infty}(43)^{\infty}}$: (**a**) trajectory in phase plane; (**b**) displacement time history; (**c**) time histories of $G^{(0,i)}$, $i = 1, 2, 3, 4$; (**e**) and (**f**) switching G-function distribution along velocity

G-function distribution along the state variable y is also depicted in Fig. 4.12e, f, which can meet all the switching conditions of the flow through the boundaries.

Second, the trajectory of chaos through two separation boundaries is shown in Fig. 4.13a, and the corresponding time history versus state variable x is plotted in Fig. 4.13b. The time histories of G-functions are illustrated in Fig. 4.13c, d, where the passable conditions of flow switchability are satisfied with Eqs. 4.15 and 4.16.

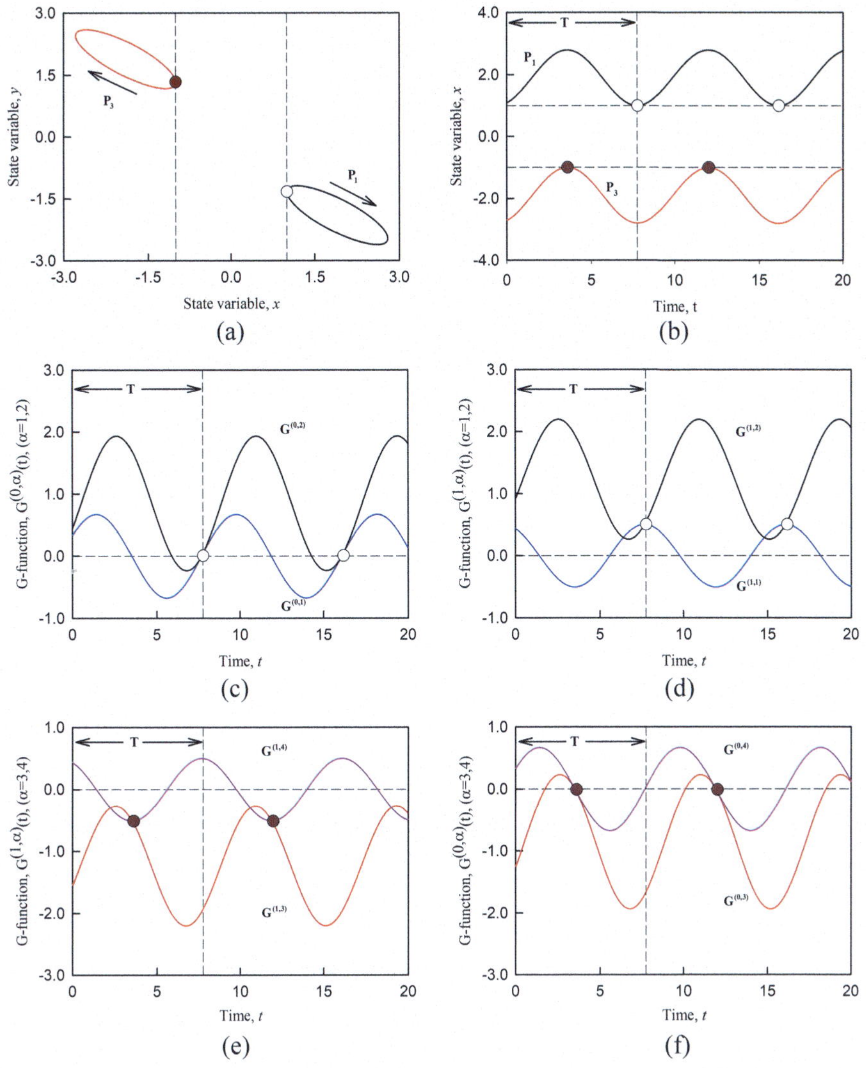

Fig. 4.14 Grazing motion of P_{21}: (**a**) coexisting trajectories in phase plane; (**b**) time history of state variables; (**c**) $G^{(0,1)}$- and $G^{(0,2)}$-function time history for right attractor; (**d**) $G^{(1,1)}$- and $G^{(1,2)}$-function time history for right attractor; (**e**) $G^{(0,3)}$- and $G^{(0,4)}$-function time history for left attractor; (**f**) $G^{(1,3)}$- and $G^{(1,4)}$-function time history for left attractor

The switching G-function distribution along state variable y is shown in Fig. 4.13e, where only the G-functions for real flows in the corresponding domains are given—the black and blue dots represent the switching points for $G^{(0,1)}$ and $G^{(0,2)}$, the pink for $G^{(0,3)}$, and the red for $G^{(0,4)}$. The switching G-function time history is given in Fig. 4.13f, which clearly illustrates how the G-functions (total forces) switch along with the switching phase. When the motion type changes, the G-functions are switched. It turns out the total forces at two separation boundaries are discontinuous.

Moreover, from the bifurcation diagram in Fig. 4.7a, the system grazing motion can be illustrated as shown in Fig. 4.14, where $\alpha = -1.448$, $\beta = -0.5$, $\sigma = 0.98$, $f = 0.15$, $\omega = 0.75$—and the initial conditions for left and right attractors are set to $t_0 = 0.0$, $x_0 = 1.091$, $y_0 = -1.199$ and $t_0 = 0.0$, $x_0 = -2.701$, $y_0 = 2.594$. The coexisting period-1 orbits are presented in Fig. 4.14a, where the trajectories for two periodic responses are tangential to the corresponding boundary, labeled by vertical dashed lines. Hereinto, the white filled symbol represents the grazing point at the boundary $x = 1$, and the coffee filled symbol represents the grazing point at the boundary $x = -1$. Two time histories of state variables are given in Fig. 4.14b, and the switching points are on the boundaries. Because two periodic attractors are in grazing motion, the unrelated G-functions are not drawn in Fig. 4.14c–f. The time histories of G-function and DG-function for right attractor are depicted in Fig. 4.14c, d. When the grazing motion happens, it can be seen that $G^{(0,1)} = 0$ and $G^{(1,1)} > 0$, which is consistent with Eq. 4.17. Similarly, the time history of G-function and DG-function for left attractor is depicted in Fig. 4.14e, f, where the system satisfies $G^{(0,4)} = 0$ and $G^{(1,4)} < 0$ at the switching time, which is consistent with Eq. 4.18.

References

Luo, A. C. J. (2008). A theory for flow switchability in discontinuous dynamical systems. *Nonlinear Analysis: Hybrid Systems, 2*(4), 1030–1061.

Luo, A. C. J. (2009). *Discontinuous dynamical systems on time-varying domains*. HEP-Springer.

Luo, A. C. J. (2012). *Discontinuous Dynamical Systems*. Higher Education Press/Springer.

Min, F. H., Zhang, W., Ji, Z. Y., & Zhang, L. (2021). Switching dynamics of a non-autonomous FitzHugh-Nagumo circuit with piecewise-linear flux-controlled memristor. *Chaos, Solitons Fractals, 152*, 111369.

Chapter 5
A Memristive Hindmarsh-Rose Neuron System

Contents

5.1 Memristor-Based Hindmarsh-Rose Circuit 57
5.2 Switchability Conditions 62
5.3 Switching Sets and Mappings 64
5.4 Bifurcation Scenarios 65
5.5 Firing Phenomenon 68
5.6 Discontinuous Dynamics 68
5.7 Hardware Implementation 73
References 75

In this chapter, the complex dynamics of non-smooth Hindmarsh-Rose (HR) neuron system with local active memristors is discussed through the theory of discontinuous dynamical system. Six mappings through two boundaries are employed to illustrate the various switching behaviors of the system. The analytical conditions of the HR system for the switching motions, such as passable motion and grazing motion, are investigated, and the extreme multistability of the improved HR neuron is revealed through switching bifurcations, parameter maps, and attraction basins. The firing activities of various modes in the HR system and the coexistence of various firing behaviors are observed by changing the initial values. Furthermore, the hardware circuit implement of the HR system is established to validate the numerical results.

5.1 Memristor-Based Hindmarsh-Rose Circuit

As in Min and Rui (2022), consider a locally active memristor model with three stable tight hysteresis loops, as shown in Fig. 5.1, which was composed of five circuit elements, including voltage-controlled memristor W with symbol term, resistor R, capacitor C, multiplier M, and operational amplifier U. This model contains

F. Min, *Discontinuous Dynamics and System Synchronization*, Nonlinear Systems and Complexity 39, https://doi.org/10.1007/978-3-031-66648-3_5

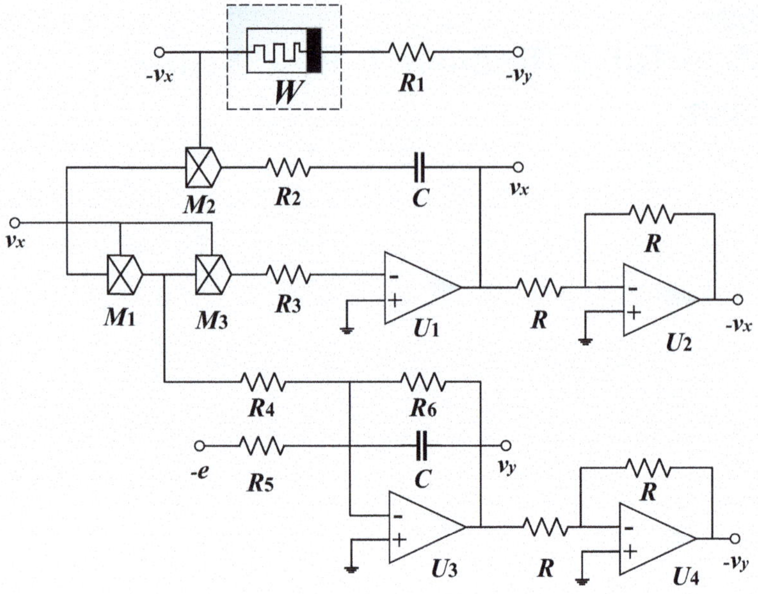

Fig. 5.1 Circuit structure diagram of HR neuron model based on memristor

connectable domains and separation boundaries, and rich dynamical behaviors and neuronal properties can be observed at the boundaries, which opens up new directions for exploring the dynamical behavior of local active memristors. In the circuit diagram of the HR model, the memristor is highlighted with the dashed line, and its equivalent circuit diagram is presented in Fig. 5.2.

With Kirchoff's law, the equivalent equation of the circuit can be described as

$$\begin{cases} C\dfrac{dv_x}{dt} = \dfrac{v_y}{R_A} + \dfrac{gv_x^2}{R_B} - \dfrac{g^2v_x^3}{R_C} + \dfrac{kv_xv_z}{R} \\ C\dfrac{dv_y}{dt} = \dfrac{e}{R_s} - \dfrac{gv_x^2}{R_D} - \dfrac{v_y}{R_F} \\ C\dfrac{dv_z}{dt} = \dfrac{R_f}{R_a}\left(\dfrac{13.5\,\mathrm{sgn}(v_z+e)}{R_s} + \dfrac{13.5\,\mathrm{sgn}(v_z-e)}{R_s} - \dfrac{v_z}{R_c}\right) + \dfrac{\beta v_x}{R} \end{cases} \tag{5.1}$$

where R_A, R_B, R_C, R_D, R_E, and R_F are all the resistors in the neuron equivalent circuit shown in Fig. 5.1, and the values are set to 10 kΩ, 333 Ω, 100 Ω, 200 Ω, 10 kΩ, and 10 kΩ, respectively. $g = 0.1$ represents gain index. In the parameters involved in Fig. 5.2, $R_s = 13.5$ kΩ and $R_f/R_a = 0.1\alpha$, α and β represent the memristor parameters, and k means the coupling strength. Moreover, the values of R and C to the above diagrams are 10 kΩ and 100 nF.

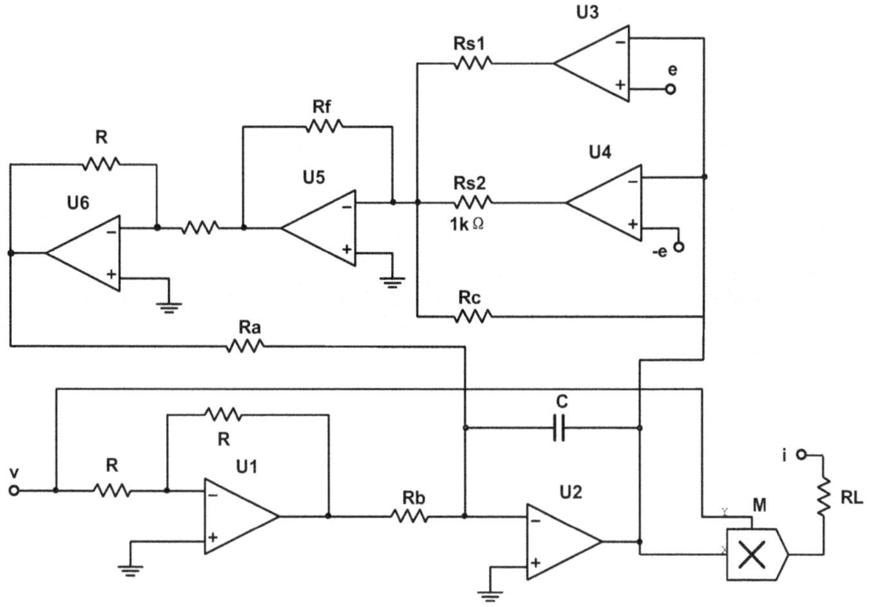

Fig. 5.2 Equivalent circuit diagram of voltage-controlled memristor *W* with symbolic function

The voltage-controlled memristor is expressed as

$$\left.\begin{array}{l} i = W(x)v = xv \\ dx/dt = F(x, v) = \alpha f(x) + \beta v \end{array}\right\} \tag{5.2}$$

where $f(x) = \text{sgn}(x + 1) + \text{sgn}(x - 1) - x$, α and β are two memristor synapse parameters. It is worth noting that the hysteresis loops of the three-segment magnetron memristor under different initial values, signal frequencies, and amplitudes are also different. When other parameters are fixed with different signal amplitude V_m, signal frequency F, and initial value $X(0)$, respectively, the hysteresis loops are depicted in Fig. 5.3. Interestingly, the three hysteresis loops of the neuron model all show an oblique "8" shape shrinking at the origin. It can be clearly observed that the lobe area is proportional to the magnitude of the signal amplitude and initial value and inversely proportional to the magnitude of the signal frequency. That is to say, the above analysis proves that the memristive properties are fully reflected in this model.

By injecting an external periodic current (AC) into 2D HR model, the mathematic expression of this memristor-based, external periodical forcing HR neural model can be derived as

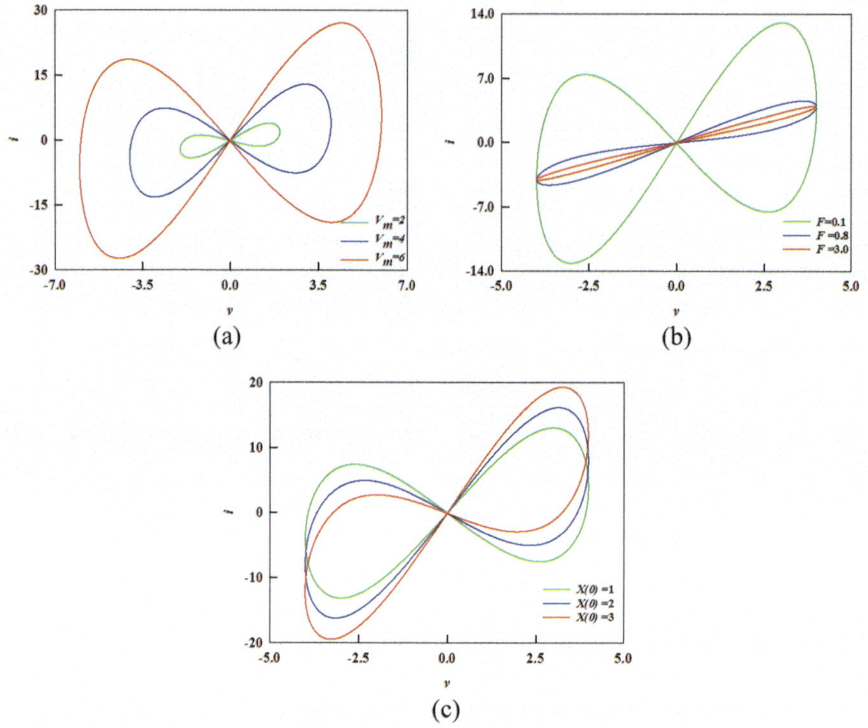

Fig. 5.3 Memristor hysteresis loops: (**a**) variation with amplitude; (**b**) variation with frequency; (**c**) variation with initial value

$$\left.\begin{aligned} \dot{x} &= y - ax^3 + bx^2 + kxz + I \\ \dot{y} &= c - dx^2 - y \\ \dot{z} &= \alpha f(z) + \beta x \end{aligned}\right\} \tag{5.3}$$

where $f(z) = \operatorname{sgn}(z+1) + \operatorname{sgn}(z-1) - z$ and AC is expressed as $I = f\cos(\omega t)$.

The state variables and vector fields of the system are defined as $\mathbf{x} = (x, y, z)^T$ and

$$\mathbf{F}(\mathbf{x}) = \begin{pmatrix} y - ax^3 + bx^2 + kxz + I \\ c - dx^2 - y \\ \alpha f(z) + \beta x \end{pmatrix} \tag{5.4}$$

where

$$f(z) = \begin{pmatrix} -2 - z, & z < -1 \\ -z, & -1 < z < 1 \\ 2 - z, & z > 1 \end{pmatrix} \tag{5.5}$$

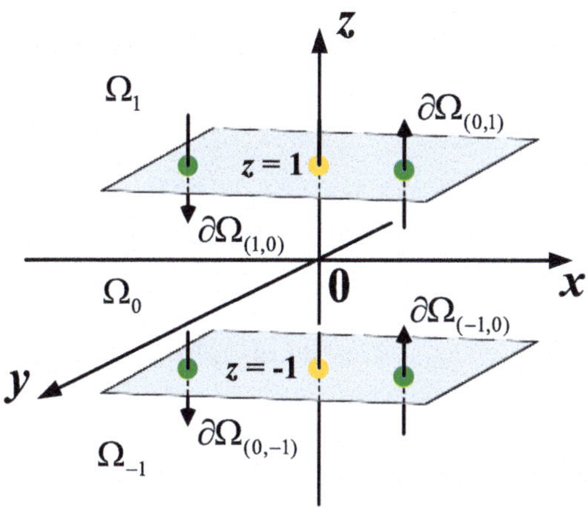

Fig. 5.4 Domains and boundaries

Because the function $f(z)$ is piecewise discontinuous, there are three connectable domains and two separated boundaries, as plotted in Fig. 5.4, where three domains are divided by

$$\left.\begin{array}{l}\Omega_{-1} = \{(x, y, z)|z < -1\} \\ \Omega_0 = \{(x, y, z)| -1 < z < 1\} \\ \Omega_1 = \{(x, y, z)|z > 1\}\end{array}\right\} \tag{5.6}$$

and the oriented boundaries $\partial\Omega_{(i,j)}$ can be expressed by

$$\left.\begin{array}{l}\partial\Omega_{(0,-1)} = \partial\Omega_{(-1,0)} = \overline{\Omega}_{-1} \cap \overline{\Omega}_0 = \{(x, y, z)|z = -1\} \\ \partial\Omega_{(0,1)} = \partial\Omega_{(1,0)} = \overline{\Omega}_1 \cap \overline{\Omega}_0 = \{(x, y, z)|z = 1\}\end{array}\right\} \tag{5.7}$$

with $\overline{\Omega}_i$ near Ω_i ($i = -1, 0, 1$).

As the total force in 5.3 is discontinuous and the flow in each domain is continuous, the motion equations of this memristive neuron system can be given by

$$\dot{\mathbf{x}} = \mathbf{F}_\lambda^{(\kappa)}(\mathbf{x}, t), \ \ (\kappa, \lambda) \in \{(-1, 0), (0, -1), (1, 0), (0, 1)\} \tag{5.8}$$

where λ and κ stand for two connectable domains, and

$$\left.\begin{array}{l}\mathbf{F}_i^{(i)}(\mathbf{x},t)=(\dot{z},F_i(\mathbf{x},t))^T\text{in }\Omega_i(i\in\{-1,0,1\})\\ \mathbf{F}_i^{(j)}(\mathbf{x},t)=(\dot{z},F_j(\mathbf{x},t))^T\\ \qquad\qquad in\ \Omega_i((i,j)\in\{(-1,0),(0,-1),(1,0),(0,1)\})\\ \mathbf{F}_1^{(1)}(\mathbf{x},t)=(x,y,1)^T or\ \mathbf{F}_{-1}^{(-1)}(\mathbf{x},t)=(x,y,-1)^T\\ \qquad\qquad on\ \partial\Omega_{(i,j)}\text{for stick,}\\ \mathbf{F}_1^{(1)}(\mathbf{x},t),\mathbf{F}_{-1}^{(-1)}(\mathbf{x},t)\in\left[\mathbf{F}_i^{(i)}(\mathbf{x},t),\mathbf{F}_j^{(j)}(\mathbf{x},t)\right]\\ \qquad\qquad on\ \partial\Omega_{(i,j)}\text{for non-stick}\end{array}\right\}\tag{5.9}$$

$\mathbf{F}_i(\mathbf{x}, t)$ is defined as the scalar force in i-domain. The vector field $\mathbf{F}_i^{(i)}(\mathbf{x},t)$ is real in i-domain and imaginary in j-domain, and $\mathbf{F}_i^{(j)}(\mathbf{x},t)$ is real in j-domain and imaginary in i-domain. $\mathbf{F}_1^{(1)}(\mathbf{x},t)$ and $\mathbf{F}_{-1}^{(-1)}(\mathbf{x},t)$ are the vector fields on boundaries ($z=1, z=-1$). The forces influenced by four oriented boundaries are given as

$$F_s(\mathbf{x},t)=\alpha^{(s)}f^{(s)}(z)+\beta^{(s)}x\ (s=1,2,3,4)\tag{5.10}$$

5.2 Switchability Conditions

As in Luo (Luo 2008, 2009, 2012), the conditions for a switching motion at the boundary of the discontinuous HR system are expressed as follows

$$\left.\begin{array}{l}n_{\partial\Omega_{(i,j)}}^T\cdot\mathbf{F}^{(i)}\left(\mathbf{x}_m^{(i)},t_{m\pm}\right)<0\\ n_{\partial\Omega_{(i,j)}}^T\cdot\mathbf{F}^{(j)}\left(\mathbf{x}_m^{(j)},t_{m\pm}\right)<0\end{array}\right\}\text{for }(i,j)\in\{(1,0),(0,-1)\}\tag{5.11}$$

$$\left.\begin{array}{l}n_{\partial\Omega_{(i,j)}}^T\cdot\mathbf{F}^{(i)}\left(\mathbf{x}_m^{(i)},t_{m\pm}\right)>0\\ n_{\partial\Omega_{(i,j)}}^T\cdot\mathbf{F}^{(j)}\left(\mathbf{x}_m^{(j)},t_{m\pm}\right)>0\end{array}\right\}\text{for }(i,j)\in\{(-1,0),(1,0)\}\tag{5.12}$$

where t_{m+} or t_{m-} is the switching moment in the switching set, which represents that the motion trajectory is just approaching to or gradually leaving away from the boundary. $t_{m\pm}=t_m\pm 0$ indicates that the motion trajectory is very close to the boundaries.

As in Luo (Luo 2008, 2009, 2012), the necessary and sufficient conditions of grazing movement tangent to the boundary are

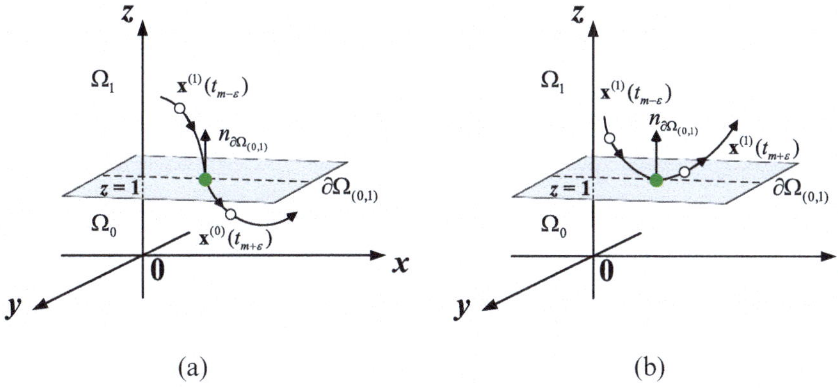

Fig. 5.5 Switching conditions: (**a**) passable flow; (**b**) grazing flow

$$\left.\begin{array}{l} n^T_{\partial\Omega_{(i,j)}} \cdot \mathbf{F}^{(i)}\left(\mathbf{x}^{(i)}_m, t_{m\pm}\right) = 0 \\ n^T_{\partial\Omega_{(i,j)}} \cdot D\mathbf{F}^{(i)}\left(\mathbf{x}^{(i)}_m, t_{m\pm}\right) > 0 \end{array}\right\} \text{for } (i,j) \in \{(1,0), (0,-1)\} \tag{5.13}$$

$$\left.\begin{array}{l} n^T_{\partial\Omega_{(i,j)}} \cdot \mathbf{F}^{(i)}\left(\mathbf{x}^{(i)}_m, t_{m\pm}\right) = 0 \\ n^T_{\partial\Omega_{(i,j)}} \cdot D\mathbf{F}^{(i)}\left(\mathbf{x}^{(i)}_m, t_{m\pm}\right) < 0 \end{array}\right\} \text{for } (i,j) \in \{(-1,0), (1,0)\} \tag{5.14}$$

in domain Ω_i, and $D\mathbf{F}^{(i)}(\mathbf{x}, t) = (F_i(\mathbf{x}, t), \nabla F^{(i)}(\mathbf{x}, t) \cdot \mathbf{F}^{(i)}(\mathbf{x}, t) + \partial_t F_\alpha(\mathbf{x}, t))^T$.

As in Luo (Luo 2008, 2009, 2012), the necessary and sufficient conditions for sliding motion disappearing from the boundary are

$$\left.\begin{array}{l} \mathbf{n}^T_{\partial\Omega_{(i,j)}} \cdot \mathbf{F}^{(i)}(\mathbf{x}_m, t_{m-}) < 0 \\ \mathbf{n}^T_{\partial\Omega_{(i,j)}} \cdot \mathbf{F}^{(j)}(\mathbf{x}_m, t_{m+}) > 0 \end{array}\right\} \text{for } n_{\partial\Omega_{(i,j)}} \to \Omega_i \tag{5.15}$$

where $(i,j) \in \{(-1,0), (0,-1), (1,0), (0,1)\}$.

There is no sliding motion of the system (5.3) at the boundaries. The passable flow and the grazing flow at the boundary $\partial\Omega_{(0,1)}$ are illustrated in Fig. 5.5, where the direction of the normal vector fields is upward.

As four oriented boundaries rely on $\varphi(x, y, z) = z \pm 1 = 0$, the normal vectors of two separation boundaries are

$$n_{\partial\Omega_{(-1,0)}} = n_{\partial\Omega_{(0,-1)}} = n_{\partial\Omega_{(1,0)}} = n_{\partial\Omega_{(0,1)}} = (0,0,1)^{\mathrm{T}} \tag{5.16}$$

As can be observed from Fig. 5.4, the normal vectors of boundaries always point to the plane $x - y$. The zero and first G-functions are

$$\left.\begin{aligned} G^{(0,s)}(\mathbf{x}_m, t_{m\pm}) &= \mathbf{n}^T_{\partial\Omega_{(i,j)}} \cdot \mathbf{F}^{(i)}(\mathbf{x}_m, t_{m\pm}) = \mathbf{F}_i(\mathbf{x}_m, t_{m\pm}), \\ G^{(1,s)}(\mathbf{x}_m, t_{m\pm}) &= \mathbf{n}^T_{\partial\Omega_{(i,j)}} \cdot D\mathbf{F}^{(i)}(\mathbf{x}_m, t_{m\pm}) = D\mathbf{F}_i(\mathbf{x}_m, t_{m\pm}) \end{aligned}\right\} \tag{5.17}$$

with $s = 1, 2, 3, 4$. The zero and first G-functions of $s = 1, 2$ are for the boundary $z = 1$, and that of $s = 3, 4$ is for $z = -1$.

5.3 Switching Sets and Mappings

The mapping structures can be used to describe various motion states and trajectories of the system at two separation boundaries. The switching set generated on boundary consists of switching points and switching time. For the system (5.3), the mapping structures and switching sets can be illustrated as shown in Fig. 5.6. The switching sets obtained from two separated boundaries ($z = 1, z = -1$) are defined as

$$\left.\begin{aligned} \Xi_1^+ &= \{(x_k, y_k, z_k, t_k) | z_k = 1^+\} \\ \Xi_1^- &= \{(x_k, y_k, z_k, t_k) | z_k = 1^-\} \\ \Xi_{-1}^+ &= \{(x_k, y_k, z_k, t_k) | z_k = -1^+\} \\ \Xi_{-1}^- &= \{(x_k, y_k, z_k, t_k) | z_k = -1^-\} \end{aligned}\right\} \tag{5.18}$$

where "+" or "−" identifies that the switching set is infinitely close to the boundary. From the above switching sets (5.18), the mapping structures can be described as

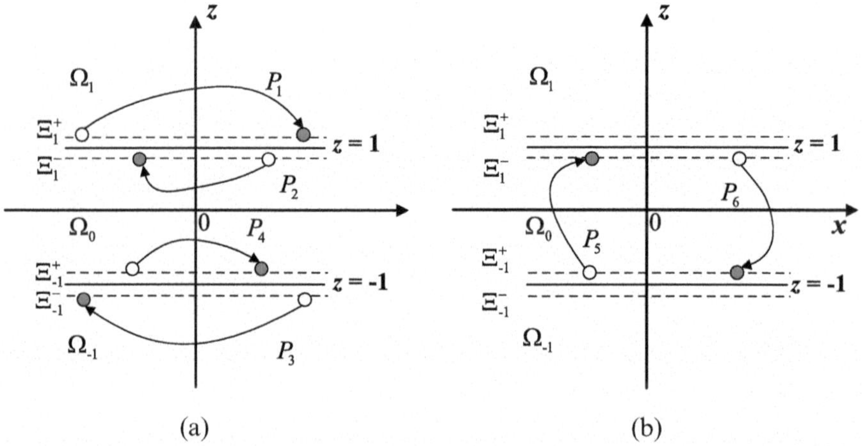

Fig. 5.6 Switching planes and mapping structures: (**a**) local mappings; (**b**) global mappings

$$\left.\begin{array}{ll} P_1 : \Xi_1^+ \to \Xi_1^+ & P_2 : \Xi_1^- \to \Xi_1^- \\ P_3 : \Xi_{-1}^- \to \Xi_{-1}^- & P_4 : \Xi_{-1}^+ \to \Xi_{-1}^+ \end{array}\right\} \quad \text{for local mappings} \tag{5.19}$$

$$P_5 : \Xi_1^- \to \Xi_{-1}^+ \qquad P_6 : \Xi_{-1}^+ \to \Xi_1^- \qquad \text{for global mappings} \tag{5.20}$$

where P_l $(l = 1, 2, 3, 4)$ is the local mapping that maps the initial condition (x_k, y_k, z_k, t_k) $(z_k \in \{(-1, 1)\})$ to the final state $(x_{k+1}, y_{k+1}, z_{k+1}, t_{k+1})$ $(z_{k+1} \in \{(-1, 1)\})$ in the same switching set, and P_l $(l = 5, 6)$, the global mapping, maps the initial condition to each other between two switching sets, finally obtains the final state.

The algebraic equations of four local mappings P_l $(l = 1, 2, 3, 4)$ are

$$\left.\begin{array}{l} g_1^{(s)} \equiv x_{k+1} - \Phi_1^{(i)}(x_k, y_k, \pm 1, t_{k+1}, t_k) = 0 \\ g_2^{(s)} \equiv y_{k+1} - \Phi_2^{(i)}(x_k, y_k, \pm 1, t_{k+1}, t_k) = 0 \\ g_3^{(s)} \equiv \pm 1 - \Phi_3^{(i)}(x_k, y_k, \pm 1, t_{k+1}, t_k) = 0 \end{array}\right\} \tag{5.21}$$

with $s = 1, 2, 3, 4$, and the algebraic equations of two global mappings P_l $(l = 5, 6)$ are

$$\left.\begin{array}{l} g_1^{(s)} \equiv x_{k+1} - \Phi_1^{(i)}(x_k, y_k, \mp 1, t_{k+1}, t_k) = 0 \\ g_2^{(s)} \equiv y_{k+1} - \Phi_2^{(i)}(x_k, y_k, \mp 1, t_{k+1}, t_k) = 0 \\ g_3^{(s)} \equiv \pm 1 - \Phi_3^{(i)}(x_k, y_k, \mp 1, t_{k+1}, t_k) = 0 \end{array}\right\} \tag{5.22}$$

where $\Phi^{(i)} = \left(\Phi_1^{(i)}, \Phi_2^{(i)}, \Phi_3^{(i)}\right)^T$.

If the system satisfies $\mathbf{x}_{k+2} = \mathbf{P}\mathbf{x}_k$ during periodic motion, the periodicity of the motion can be described as

$$x_{k+2} = x_k, t_{k+2} = t_k + T \tag{5.23}$$

where $\mathbf{x}_k = (x_k, y_k, \omega t_k)^T$ and T represents one period of the periodic motion. By using the above analysis method, more complex or even chaotic motion trajectories can be analyzed.

5.4 Bifurcation Scenarios

The system parameters are set as $b = 3$, $c = 1$, $d = 5$, $k = 0.9$, the amplitude and frequency of the external excitation are $f = 0.1$, $\omega = 1.0$, the memristive parameters are $\alpha = 0.1$, $\beta = 0.8$, and the initial conditions are $x_0 = 0.1$, $y_0 = 0.1$, $z_0 = 0.1$, $t_0 = 0.0$.

First, for a increasing from 0.5 to 2 with $\Delta a = 0.0015$, the bifurcation diagram is shown in Fig. 5.7a, where the mapping structure of different periodic motions is

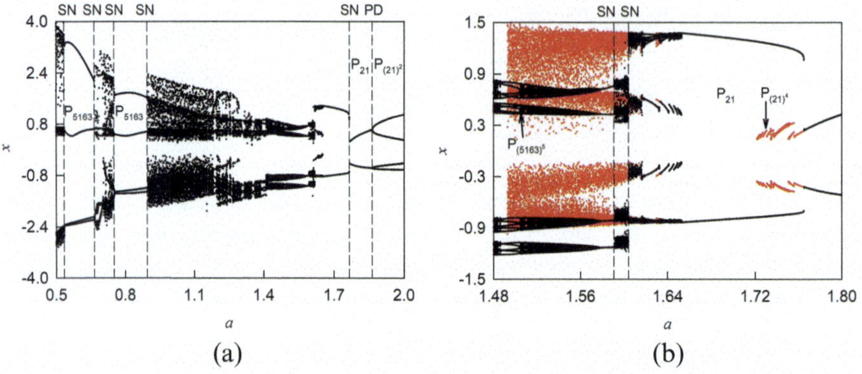

Fig. 5.7 Bifurcation diagrams with system parameter a: (**a**) global bifurcation diagram of $a \in [0.5, 2.0]$; (**b**) local coexistence bifurcation diagram of $a \in [1.48, 1.80]$

Table 5.1 Mapping structures of variable a and the corresponding ranges

Ranges of a	Coexisting mapping structures
(1.492, 1.509)	$P_{(5163)^5}$ and $P_{(5163)^\infty}$
(1.509, 1.616)	$P_{(5163)^\infty}$ and $P_{(5163)^\infty}$
(1.616, 1.721)	P_{21} and no switching motion
(1.721, 1.766)	P_{21} and $P_{(21)^4}$

marked in the corresponding areas. For the first stage, a increasing from $a = 0.5$, the system is in a chaotic state. From $a = 0.534$, the system steps into the saddle-node bifurcation (SN), the current mapping structure is P_{5163}, and its trajectory moves across the boundaries $z = 1$ and $z = -1$ each for once. A series of saddle-node bifurcations happen at $a = 0.664$, $a = 0.75$, $a = 0.896$, the system interacts with periodic and chaotic motions. After that, the system (5.3) enters a long-periodic chaotic state, until $a = 1.766$, the system undergoes saddle-node bifurcation and turns into periodic motion, its mapping structure is P_{21}. For $a = 1.864$, the system undergoes a period-doubling bifurcation (PD), the mapping structure changes to $P_{(21)^2}$, and the motion trajectory appears to switch across the boundary $z = 1$ twice, without intersecting with the boundary $z = -1$. For further observation of the coexistence, the interval of a is zoomed into [1.48,1.8] with $\Delta a = 0.00032$. By increasing or decreasing a, respectively, the coexistence bifurcation can be obtained as shown in Fig. 5.7b, where the corresponding parameter ranges and mapping structures are listed in Table 5.1. For $a \in [1.492, 1.509]$, the multiple periodic and chaotic motions with $P_{(5163)^5}$ and $P_{(5163)^\infty}$ coexist. For $a \in [1.509, 1.616]$, the coexistence of chaos and chaos can be observed, which switches across both separation boundaries $z = 1$ and $z = -1$, with $P_{(5163)^\infty}$ and $P_{(5163)^\infty}$. For $a \in [1.616, 1.721]$, the reduced a has no trajectory on the bifurcation graph, which means the phase plane's trajectory does not intersect with the separation boundaries.

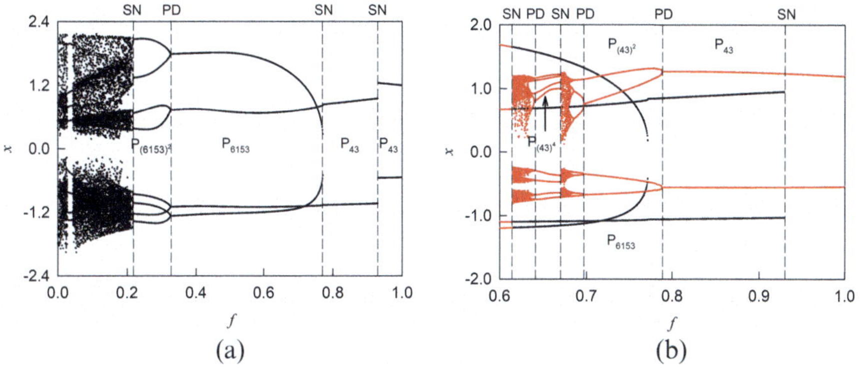

Fig. 5.8 Boundary bifurcation diagrams with amplitude f: (**a**) global bifurcation diagram of f; (**b**) local coexistence bifurcation diagram of f

However, the enlarged a (black dots) is now in period-1 of P_{21}, which only crosses the boundary $z = 1$. For $a \in [1.721, 1.766]$, two kinds of periods with mapping structures P_{21} and $P_{(21)^4}$ coexist.

Moreover, the amplitude of the periodic external excitation also has a great influence on the system switching motion. Set the system parameters $a = 1$, $b = 3$, $c = 1$, $d = 5$, $k = 0.9$, the frequency of the external excitation $\omega = 1.0$, the memristive parameters $\alpha = 0.1$, $\beta = 0.8$, and the initial conditions $x_0 = 0.0$, $y_0 = 0.0$, $z_0 = 0.0$, $t_0 = 0.0$. For the amplitude f increasing from 0 to 1 with $\Delta f = 0.001$, the bifurcation diagram is obtained as shown in Fig. 5.8a. It can be observed that the chaotic motion occurs only when $f \in (0.0, 0.218)$ is small, where a periodic window appears at $[0.0251, 0.0421]$. For $f = 0.218$, the system undergoes a saddle-node bifurcation (SN) turning into periodic motion, its mapping structure is $P_{(5163)^2}$, and its corresponding phase trajectory crosses two separation boundaries. For $f = 0.328$, a reverse period-doubling bifurcation takes place, and the periodic response of the system is P_{5163}. For $f = 0.789$, the system encounters the saddle-node bifurcation again, its mapping structures are transformed from P_{5163} into P_{21}, and its motion trajectory changes from switching two boundaries to just crossing the boundary $z = 1$. For $f = 0.929$, the system has a saddle-node bifurcation, and its mapping structure is still P_{21}, but two bifurcation routes are different. The amplitude f is zoomed into $f \in [0.6, 1.0]$ to observe the coexistence of the HR neuron system varying with the amplitude f at $\Delta f = \pm 0.0004$, which frequently obtains the coexisting bifurcation diagram as shown in Fig. 5.8b. The range of coexistence with mapping structures is listed in Table 5.2. For $f \in (0.6143, 0.641) \cup (0.67, 0.697)$, the chaotic and periodic motions coexist with the mapping structure P_{5163} for switching across two separation boundaries $z = 1$ and $z = -1$ via the global mapping P_5 and P_6 and the mapping structure $P_{(43)^\infty}$ for traversing again around the boundary $z = 1$. For $f \in (0.641, 0.67)$, the periodic coexistence with the mapping structures P_{5163} and $P_{(43)^4}$ can be presented. For $f \in (0.697, 0.789)$, the coexistence

Table 5.2 Coexisting mapping structures of the amplitude f

Ranges of f	Coexisting mapping structures
$(0.6143, 0.641) \cup (0.67, 0.697)$	P_{5163} and $P_{(43)^{\infty}}$
$(0.641, 0.67)$	P_{5163} and $P_{(43)^4}$
$(0.697, 0.789)$	P_{5163} and $P_{(43)^2}$
$(0.789, 0.93)$	P_{43} and P_{43}

of different periodic orbits P_{5163} and $P_{(43)^2}$ can be also observed. For $f \in (0.789, 0.93)$, the coexisting attractors are also illustrated with the mapping structures $P_{(43)^4}$ and $P_{(43)^4}$.

5.5 Firing Phenomenon

To investigate the system dynamic behavior affected by the external stimulus currents, its amplitude f is used as the bifurcation parameter that has been discussed in the previous section, where the relative bifurcation diagram is illustrated in Fig. 5.8a. The phase diagrams and corresponding time sequences of the membrane action potential of the HR system (5.3) under different external excitation currents with $I = f\cos(\omega t)$ are displayed in Fig. 5.9. The four firing activities, such as chaotic bursting firing with $f = 0.1$, period-8 bursting firing with $f = 0.27$, period-3 bursting firing with $f = 0.5$, and period-1 spiking firing with $f = 0.96$, can be observed, which means that the system firing patterns are quite different with different f. For f increasing, the peak value and the discharge period decrease, which can be observed by the corresponding phase diagram trajectory.

5.6 Discontinuous Dynamics

In this part, the switching behaviors of the coexisting periodic and chaotic motion at the separation boundaries of HR system in Eq. 5.3 will be illustrated through the theory of discontinuous dynamic system with parameters $a = 1, b = 3, c = 1, d = 5, k = 0.9, f = 0.1, \alpha = 0.1, \beta = 0.8576$, respectively.

First of all, the periodic motion with mapping structure P_{5163} with the above parameter is depicted in Fig. 5.10, with the initial conditions $t_0 = 0.0$, $x_0 = -0.4915$, $y_0 = -2.615$, $z_0 = -4.5106$, where the solid points are switching points and gray stands for the starting point of the periodic motion, white for the motion trajectory from Ω_0 to Ω_1, green from Ω_1 to Ω_0, orange from Ω_{-1} to Ω_0, and brown from Ω_0 to Ω_{-1}. Comparing Fig. 5.10a, b, it can be seen that the phase plane's trajectory starts from the gray point and then jumps from the switching set Ξ_{-1} to Ξ_1

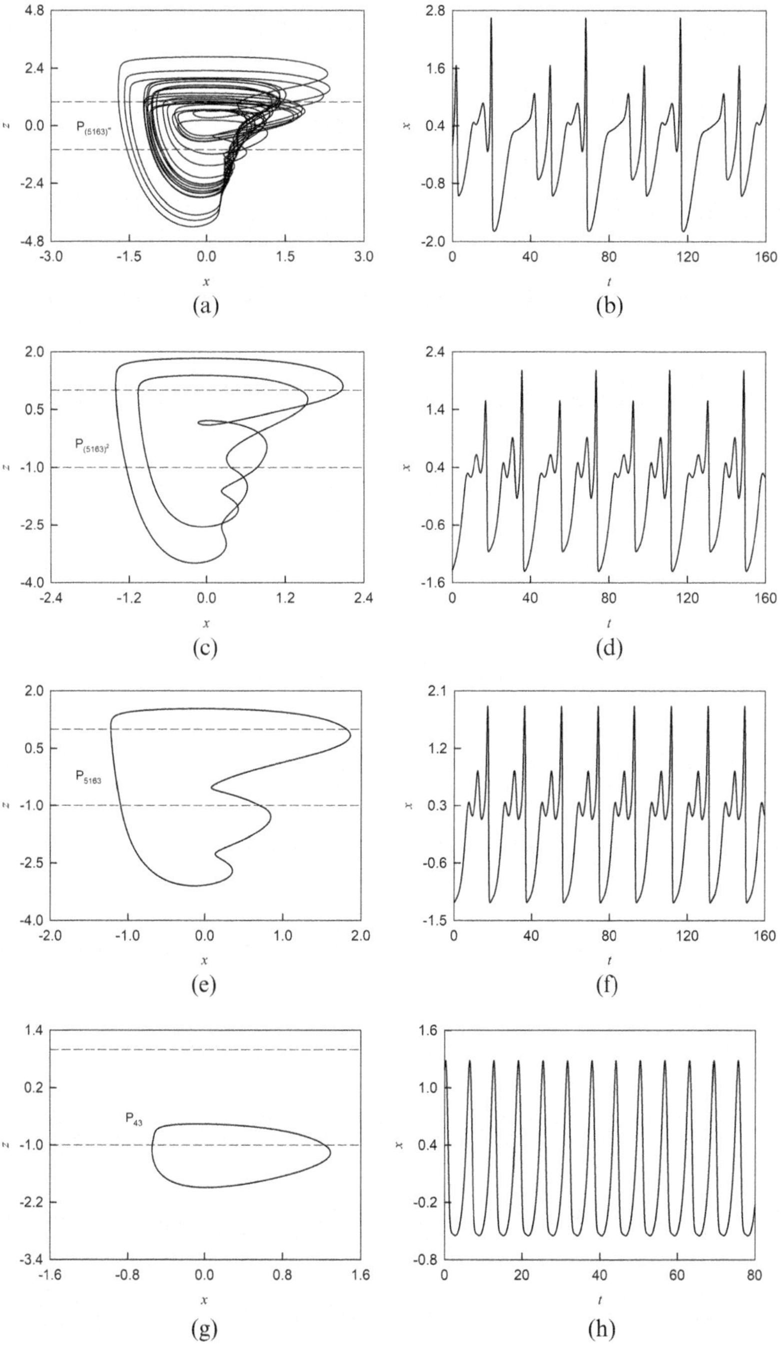

Fig. 5.9 Phase plane and time histories of firing patterns : (**a**) and (**b**) chaotic bursting firing with $f = 0.1$; (**c**) and (**d**) period-8 bursting firing with $f = 0.27$; (**e**) and (**f**) period-3 bursting firing with $f = 0.5$; (**g**) and (**h**) period-1 spiking firing with $f = 0.96$

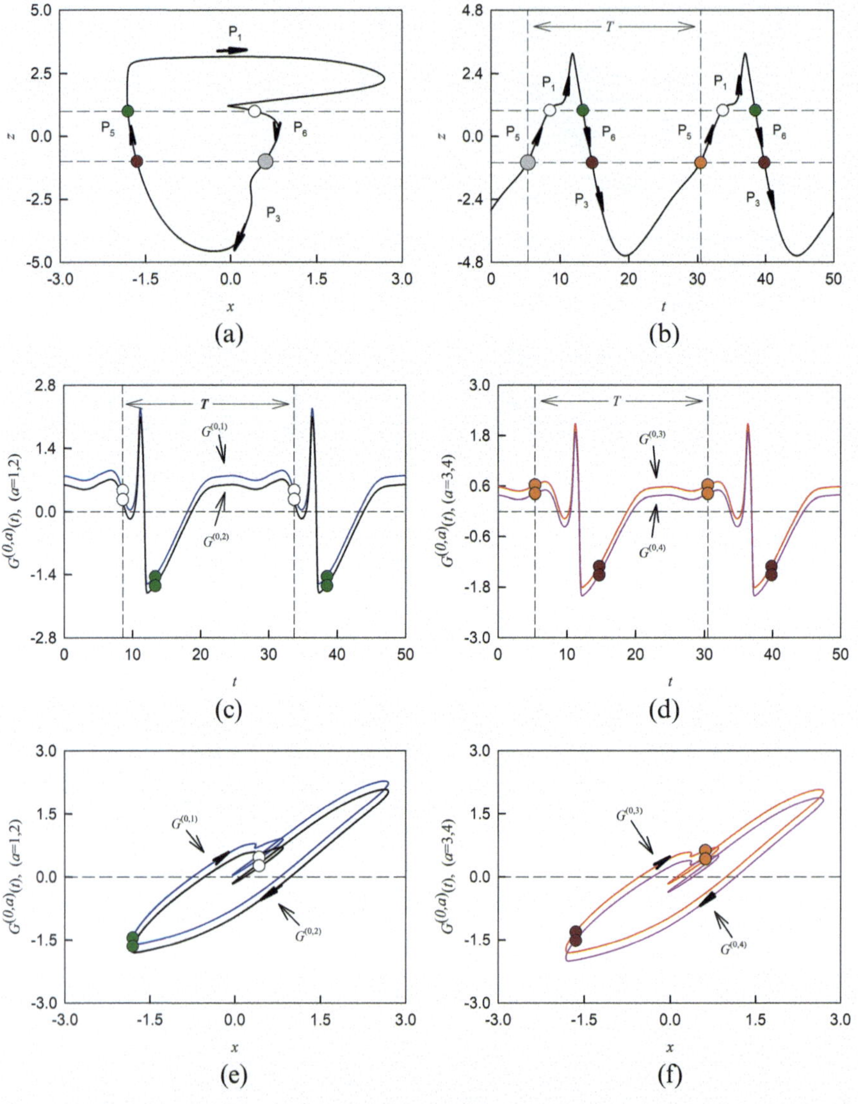

Fig. 5.10 Periodic motion of P_{5163}: (**a**) phase plane; (**b**) time sequence diagram on the z-axis; (**c**) time sequence diagram of $G^{(0,1)}$- and $G^{(0,2)}$-function; (**d**) time sequence diagram of $G^{(0,3)}$- and $G^{(0,4)}$-function; (**e**) $G^{(0,1)}$- and $G^{(0,2)}$-function distribution along x-axis; (**f**) $G^{(0,3)}$- and $G^{(0,4)}$-function distribution along x-axis

via the global mapping P_5, reaching the white point on the boundary $z = 1$. After the above motion, the trajectory turns from Ω_1 to Ω_0 through the local mapping P_1 to the green switching point, which subsequently jumps from the switching set Ξ_1 to Ξ_{-1} under the global mapping P_6, returning to the coffee point on the boundary $z = -1$. And then the trajectory returns to the start via the local mapping P_6, representing the trajectory completing the switching movement in a period. The time history diagram of the G-functions is shown in Fig. 5.10c, d, where G-functions can be judged positive or negative according to the switching points in different situations, which are consistent with the conditions (5.11) and (5.12) analyzed before. From Ω_{-1} to Ω_0, the switching point (orange) is above $G = 0$, indicating $G^{(0,3)} > 0$, $G^{(0,4)} > 0$. From Ω_0 to Ω_1, the switching point (white) is also above $G = 0$, indicating $G^{(0,1)} > 0$ and $G^{(0,2)} > 0$. From Ω_0 to Ω_{-1}, the switching point (green) is below $G = 0$, indicating $G^{(0,1)} < 0$, $G^{(0,2)} < 0$. From Ω_1 to Ω_0, the switching point (coffee) is below $G = 0$, indicating $G^{(0,3)} < 0$, $G^{(0,4)} < 0$. The distribution diagrams of the G-function on the x-axis are illustrated in Fig. 5.10e, f, where the distribution of the switching point is consistent with the description above, and the switching point at the x-axis is linked with that in phase plane when the periodic attractor crosses the boundary, which mutually verifies the boundary motion conditions.

Moreover, changing the initial conditions to $t_0 = 0.0$, $x_0 = 0.4236$, $y_0 = 0.3051$, $z_0 = -2.1821$, the system will present a chaotic motion with the mapping structure $P_{(5163)^\infty}$. The chaotic phase plane is shown in Fig. 5.11a, where only the first 2000 sets of data are saved to facilitate the observation of the boundary characteristics of the chaotic motion. The corresponding time history diagrams are shown in Fig. 5.11b, and the time history diagrams of the G-function are depicted in Fig. 5.12c, d, where the blue, black, red, and pink lines, respectively, represent $G^{(0,1)}$, $G^{(0,2)}$, $G^{(0,3)}$, and $G^{(0,4)}$, the situations of which can be judged by the switching movement and which are also consistent with the conditions 11 and 12: (1) from Ω_{-1} to Ω_0, $G^{(0,3)} > 0$ and $G^{(0,4)} > 0$; (2) from Ω_0 to Ω_1, $G^{(0,1)} > 0$ and $G^{(0,2)} > 0$; (3) from Ω_0 to Ω_{-1}, $G^{(0,1)} < 0$ and $G^{(0,2)} < 0$; (4) from Ω_1 to Ω_0, $G^{(0,3)} < 0$ and $G^{(0,4)} < 0$.

The time history diagram of the real flow in HR neuron is simulated as shown in Fig. 5.11e, where the trajectory of the phase plane, affected by the real flows in corresponding regions, travels along two separation boundaries in three connectable regions. Via the switching movement at the boundaries, it can be clearly observed that the switching points meet the conditions for switching movement.

The trajectory mapping of the G-function at the switching boundary is shown in Fig. 5.11f, where the switching points in one period can be observed. In a period, all the switching points are discrete points and converged into several curves on the mapping plane, which illustrates that the system is in chaos. $G^{(0,1)}$ and $G^{(0,3)}$ is within the positive part of the vertical axis, indicating that these switching points are greater than 0. $G^{(0,2)}$ and $G^{(0,4)}$ is within the negative part of the vertical axis, indicating that these switching points are less than 0 and satisfy the analysis conditions (5.11) and (5.12).

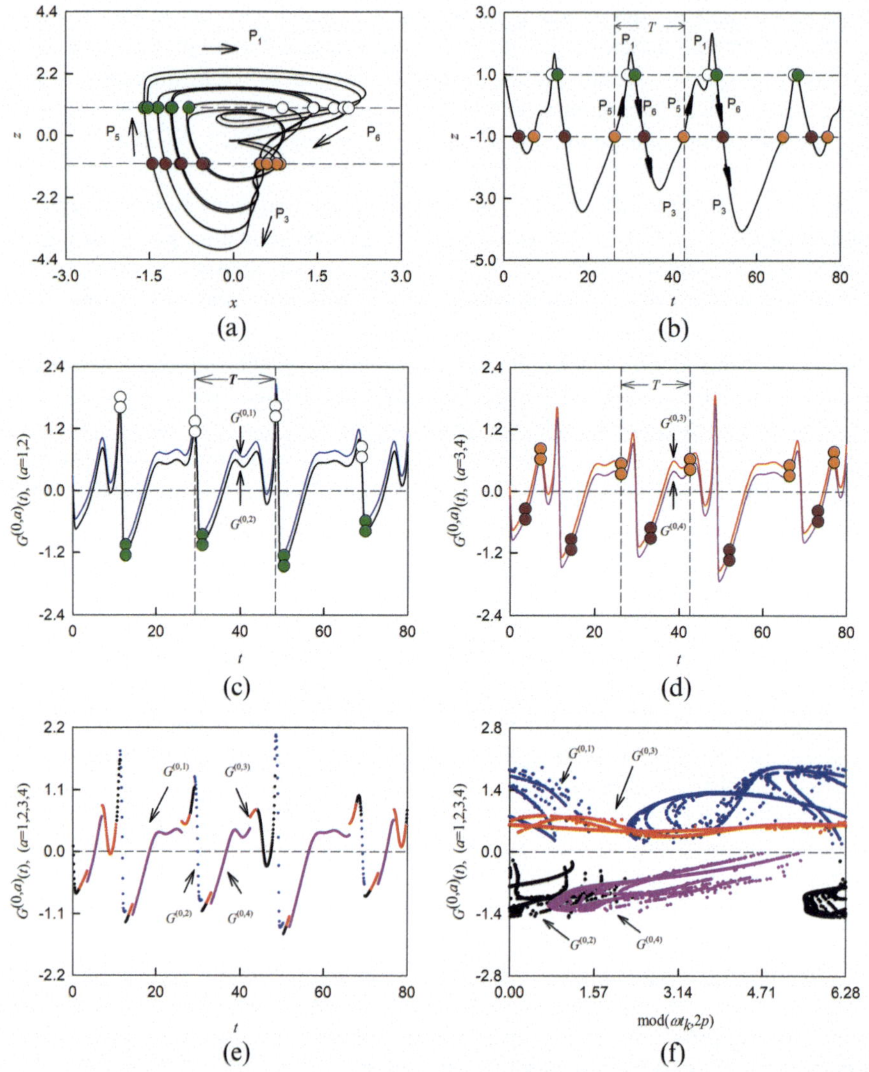

Fig. 5.11 Chaotic motion of $P_{(5163)^{\infty}}$: (**a**) phase plane; (**b**) time sequence diagram on the y-axis; (**c**) time sequence diagram of $G^{(0,1)}$- and $G^{(0,2)}$-function; (**d**) time sequence diagram of $G^{(0,3)}$- and $G^{(0,4)}$-function; (**e**) real flow timing diagram; (**f**) G-function trajectory mapping on the switching boundary

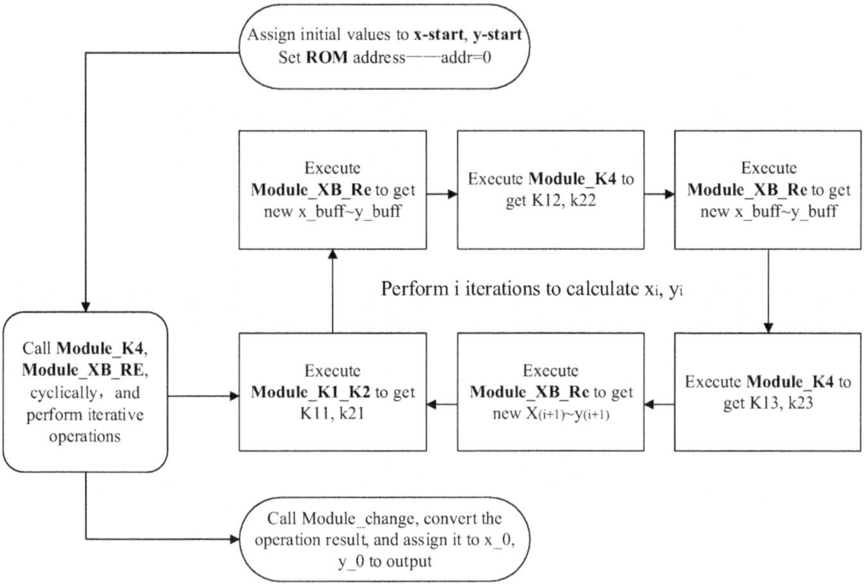

Fig. 5.12 Operation flowchart of Module_HR

5.7 Hardware Implementation

In this paper, field programmable gate array (FPGA) is used to conduct the circuit experiments on the improved HR memristive neuron model that is discretized, where FPGA is a semi-custom modular integrated circuit that can be reconfigured by designers after fabrication using a hardware description language such as Verilog or VHDL. The FPGA design scheme starts with discretizing the system (5.2) and then simulates and tests the discrete mathematical model via MATLAB. After the basic module is designed with Verilog HDL language, the program is compiled and then burned to the FPGA development board. Lastly, the dynamic behaviors can be observed and captured by the oscilloscope. The coexisting attractors of this memristive neuron system will be observed and captured through an oscilloscope.

Here, second order Runge-Kutta method is used to discretize the system equation (5.2), and the discretization equation of the HR neuron system can be given by

$$\begin{cases} x_{i+1} = T^*(K_{11} + K_{12})/2 \\ y_{i+1} = T^*(K_{21} + K_{22})/2 \\ z_{i+1} = T^*(K_{31} + K_{32})/2 \end{cases} \tag{5.24}$$

where the iteration parameter K_{11}~K_{31}, K_{12}~K_{32}, K_{13}~K_{33} and K_{14}~K_{34} are

$$\begin{cases} K_{11} = y_i - ax_i^3 + bx_i^2 + kx_i z_i + f\cos(\omega t) \\ K_{21} = c - dx_i^2 - y_i \\ K_{31} = \alpha(\mathrm{sgn}(z_i + 1) + \mathrm{sgn}(z_i - 1) - z_i) + \beta x_i \end{cases} \tag{5.25}$$

$$\begin{cases} K_{12} = y_i - a(x_i + T * K_{11})^3 + b(x_i + T * K_{11})^2 + k(x_i + T^* K_{11})(z_i + T^* K_{31}) \\ +f\cos(\omega(t + T)) \\ K_{22} = c - d(x_i + T * K_{11})^2 - (y_i + T^* K_{21}) \\ K_{32} = \alpha(\mathrm{sgn}((z_i + T^* K_{31}) + 1) + \mathrm{sgn}((z_i + T^* K_{31}) - 1) - (z_i + T^* K_{31})) \\ +\beta(x_i + T^* K_{11}) \end{cases} \tag{5.26}$$

$$\begin{cases} K_{13} = y_i - a(x_i + T * K_{12})^3 + b(x_i + T * K_{12})^2 + k(x_i + T^* K_{12})(z_i + T^* K_{32}) \\ +f\cos(\omega(t + T)) \\ K_{23} = c - d(x_i + T * K_{12})^2 - (y_i + T^* K_{22}) \\ K_{33} = \alpha(\mathrm{sgn}((z_i + T^* K_{32}) + 1) + \mathrm{sgn}((z_i + T^* K_{32}) - 1) - (z_i + T^* K_{32})) \\ +\beta(x_i + T^* K_{12}) \end{cases} \tag{5.27}$$

According to 5.24, the whole program is designed into four modules, including a top-level module (Module_HR) and three bottom-level modules (Module_K4, Module_XB_RE, Module_change). Module_HR calls Module_K4 and Module_XB_RE in sequence to realize the differential equation system. The operation flowchart is shown as in Fig. 5.12, and each module can play various operations role.

The initial conditions, the system, and memristor parameters are set the same as that of the previous simulation, the results of which are illustrated in Figs. 5.11 and 5.12. Carrying out the FPGA implementation by changing f or initial conditions, the system dynamic behavior can be observed and captured via the oscilloscope, as shown in Fig. 5.13. The top half of the figure shows the two-channel time series, and the bottom half shows the phase diagram. By comparing the result of experiment with those of simulations, the consistency of the system firing activities can also be observed. Clearly, periodic motion of HR model with P_{5163} crosses two boundaries at the same time, and chaotic motion with mapping structure $P_{(5163)^\infty}$ of the system performs random chaotic traversal motion.

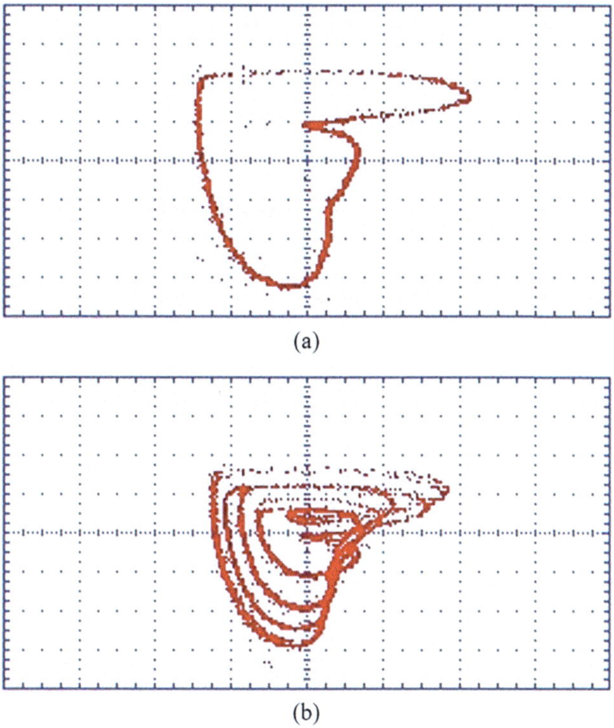

Fig. 5.13 Hardware circuit phase diagram and time sequence diagram, CH1=1v/div, CH2=1V/div: (a) periodic motion with P_{5163} and $x_0 = -0.4915$, $y_0 = -2.615$, $z_0 = -4.5106$; (b) chaotic motion with $P_{(5163)^\infty}$ and $x_0 = 0.4236$, $y_0 = 0.3051$, $z_0 = -2.1821$

References

Luo, A. C. J. (2008). A theory for flow switchability in discontinuous dynamical systems. *Nonlinear Analysis: Hybrid Systems, 2*(4), 1030–1061.

Luo, A. C. J. (2009). *Discontinuous Dynamical Systems on Time-varying Domains*. HEP-Springer.

Luo, A. C. J. (2012). *Discontinuous Dynamical Systems*. Higher Education Press/Springer.

Min, F. H. and Rui, Z. (2022), Boundary dynamics of a non-smooth memristive Hindmarsh–Rose neuron system, *Chaos,* 32, pp.103117.

Chapter 6
Dynamical System Interaction Theory

Contents

6.1 Introduction to System Interactions 77
6.2 An Interaction Theory 82
6.3 Edge Interaction Theory 88
References 91

Dynamical system synchronization is an application of system interactions. From Luo (2009, 2012), the concept of interaction between two dynamical systems is reviewed. An interaction condition of two dynamical systems is considered as a separation boundary, and such a separation boundary is time-varying. Luo (2009, 2012) developed the system interaction theory from the theory for the switchability and attractivity of edge flows to the specific edges in discontinuous dynamical systems.

6.1 Introduction to System Interactions

From Luo (2009, 2012), basic concepts of the dynamical system interactions will be reviewed, and the discontinuous theory of the interaction of two dynamical systems is stated as follows.

From Luo (2009, 2012), the interaction constraints will be introduced first.

Definition 6.1 Two dynamical systems are defined by

$$\dot{\mathbf{y}} = \mathbf{F}(\mathbf{y}, t, \mathbf{p}) \in \mathcal{R}^n \text{ and } \dot{\mathbf{x}} = \boldsymbol{\mathcal{F}}(\mathbf{x}, t, \mathbf{q}) \in \mathcal{R}^m \tag{6.1}$$

If two flows $\mathbf{x}(t)$ and $\mathbf{y}(t)$ of the two systems in Eq. 6.1 satisfy

F. Min, *Discontinuous Dynamics and System Synchronization*, Nonlinear Systems and Complexity 39, https://doi.org/10.1007/978-3-031-66648-3_6

$$\varphi(\mathbf{x}(t),\mathbf{y}(t),t,\boldsymbol{\lambda})=0,\ \ \boldsymbol{\lambda}\in\mathcal{R}^{n_0} \tag{6.2}$$

then the two systems are called to be interacted (or constrained) under such a condition at time t.

In fact, the interaction of two dynamical systems can occur under many constraints instead of Eq. 6.2, and the following definition is given in Luo (2009, 2012).

Definition 6.2 Consider l-nonidentical functions of $\varphi_j(\mathbf{x}(t),\mathbf{y}(t),t,\boldsymbol{\lambda}_j)$ ($j\in\mathcal{L}$ and $\mathcal{L}=\{1,2,\cdots,l\}$). If two flows $\mathbf{x}(t)$ and $\mathbf{y}(t)$ of two systems in Eq. 6.1 satisfy for time t

$$\varphi_j(\mathbf{x}(t),\mathbf{y}(t),t,\boldsymbol{\lambda}_j)=0 \text{ for } \boldsymbol{\lambda}_j\in\mathcal{R}^{n_j} \text{ and } j\in\mathcal{L}, \tag{6.3}$$

then two systems in Eq. 6.1 are called to be constrained under the jth-condition at time t.

For the foregoing definition, two dynamical systems in Eq. 6.1 possess l-conditions for interactions (or constraints). Thus, the l-separation boundaries relative to the interaction divide the corresponding phase space into many sub-domains for the two dynamical systems, and these sub-domains change with time.

From Luo (2009, 2012), the boundary is determined by the jth-condition in Eq. 6.3, and the corresponding domain in phase space is divided into two domains $\mho_{(\alpha_j,j)}$ and $\underline{\mho}_{(\alpha_j,j)}$ ($\alpha_j=1,\,2$) for two dynamical systems in Eq. 6.1, respectively. From the definition in Luo (2009, 2012), we have the following definition.

Definition 6.3 On the open sub-domain $\mho_{(\alpha_j,j)}$, there is a $C^{r_{\alpha_j}}$-continuous system ($r_{\alpha_j}\geq 1$) in a form of

$$\begin{aligned}
&\dot{\mathbf{y}}^{(\alpha_j,j)}=\mathbf{F}^{(\alpha_j,j)}\left(\mathbf{y}^{(\alpha_j,j)},t,\mathbf{p}^{(\alpha_j,j)}\right)\in\mathcal{R}^n,\\
&\mathbf{y}^{(\alpha_j,j)}=\left(y_1^{(\alpha_j,j)},y_2^{(\alpha_j,j)},\cdots,y_n^{(\alpha_j,j)}\right)^{\mathrm{T}}\in\mho_{(\alpha_j,j)}.
\end{aligned} \tag{6.4}$$

In a sub-domain $\mho_{(\alpha_j,j)}$, the vector field of $\mathbf{F}^{(\alpha_j,j)}\left(\mathbf{y}^{(\alpha_j,j)},t,\mathbf{p}^{(\alpha_j,j)}\right)$ with parameter $\mathbf{p}^{(\alpha_j,j)}=\left(p_1^{(\alpha_j,j)},p_2^{(\alpha_j,j)},\cdots,p_k^{(\alpha_j,j)}\right)^{\mathrm{T}}\in\mathcal{R}^{k_j}$ is $C^{r_{\alpha_j}}$-continuous ($r_{\alpha_j}\geq 1$) for a state vector $\mathbf{x}^{(\alpha_j,j)}$ and time t. The continuous flow of the first dynamical system in Eq. 6.1 (i.e., $\mathbf{y}^{(\alpha_j,j)}(t)=\boldsymbol{\Phi}^{(\alpha_j,j)}\left(\mathbf{y}^{(\alpha_j,j)}(t_0),t,\mathbf{p}^{(\alpha_j,j)}\right)$) is $C^{r_{\alpha_j}+1}$-continuous for time t with initial condition $\mathbf{y}^{(\alpha_j,j)}(t_0)=\boldsymbol{\Phi}^{(\alpha_j,j)}\left(\mathbf{y}^{(\alpha_j,j)}(t_0),t_0,\mathbf{p}^{(\alpha_j,j)}\right)$.

The corresponding boundary relative to the jth non-connected interaction is defined in Luo (2009, 2012) as follows.

Definition 6.4 The interaction boundary in n-dimensional phase space for the first dynamical system in Eq. 6.1 under the jth interaction condition in Eq. 6.3 is defined as

$$
\begin{aligned}
S_{(\alpha_j\beta_j,j)} &\equiv \partial \mho_{(\alpha_j\beta_j,j)} = \overline{\mho}_{(\alpha_j,j)} \cap \overline{\mho}_{(\alpha_j,j)} \\
&= \left\{ \mathbf{y}^{(0,j)} \left| \begin{array}{l} \varphi_j\left(\mathbf{x}^{(0,j)}, \mathbf{y}^{(0,j)}, t, \boldsymbol{\lambda}_j\right) = 0, \\ \varphi_j \text{ is } C^{r_j}\text{-continuous } \left(r_j \geq 1\right) \end{array} \right. \right\} \subset \mathscr{R}^{n-1}.
\end{aligned} \tag{6.5}
$$

Similarly, the discontinuous system of the second system in Eq. 6.1, caused by the interaction at the jth-condition in Eq. 6.3, is described in Luo (2009, 2012).

Definition 6.5 On the α_jth ($\alpha_j = 1, 2$) -open sub-domain $\underline{\mho}_{(\alpha_j,j)}$, there is a $C^{s_{\alpha_j}}$-continuous system ($s_{\alpha_j} \geq 1$) in a form of

$$
\begin{aligned}
\dot{\mathbf{x}}^{(\alpha_j,j)} &= \mathcal{F}^{(\alpha_j,j)}\left(\mathbf{x}^{(\alpha_j,j)}, t, \mathbf{q}^{(\alpha_j,j)}\right) \in \mathscr{R}^m, \\
\mathbf{x}^{(\alpha_j,j)} &= \left(x_1^{(\alpha_j,j)}, x_2^{(\alpha_j,j)}, \cdots, x_m^{(\alpha_j,j)}\right)^{\mathrm{T}} \in \underline{\mho}_{(\alpha_j,j)}.
\end{aligned} \tag{6.6}
$$

In a sub-domain $\underline{\mho}_{(\alpha_j,j)}$, the vector field of $\mathcal{F}^{(\alpha_j,j)}\left(\mathbf{x}^{(\alpha_j,j)}, t, \mathbf{q}^{(\alpha_j,j)}\right)$ with parameter of $\mathbf{q}^{(\alpha_j,j)} = \left(q_1^{(\alpha_j,j)}, q_2^{(\alpha_j,j)}, \cdots, q_m^{(\alpha_j,j)}\right)^{\mathrm{T}} \in \mathscr{R}^{k_j}$ is $C^{s_{\alpha_j}}$-continuous ($s_{\alpha_j} \geq 1$) in a state vector $\mathbf{x}^{(\alpha_j,j)}$ and for time t. The continuous flow of the second dynamical systems in Eq. 6.1 (i.e.,$\mathbf{x}^{(\alpha_j,j)}(t) = \underline{\boldsymbol{\Phi}}^{(\alpha_j,j)}\left(\mathbf{x}^{(\alpha_j,j)}(t_0), t, \mathbf{q}^{(\alpha_j,j)}\right)$) is $C^{s_{\alpha_j}+1}$-continuous for time t.

From Luo (2009, 2012), the corresponding boundary relative to non-connected interaction is defined for the second dynamical system in Eq. 6.1.

Definition 6.6 The boundary in m-dimensional phase space for the second dynamical system in Eq. 6.1 is defined as

$$
\begin{aligned}
\underline{S}_{(\alpha_j\beta_j,j)} &\equiv \partial \underline{\mho}_{(\alpha_j\beta_j,j)} = \overline{\underline{\mho}}_{(\alpha_j,j)} \cap \overline{\underline{\mho}}_{(\beta_j,j)} \\
&= \left\{ \mathbf{x}^{(0,j)} \left| \begin{array}{l} \varphi_j\left(\mathbf{x}^{(0,j)}, \mathbf{y}^{(0,j)}, t, \boldsymbol{\lambda}_j\right) = 0, \\ \varphi_j \text{ is } C^{r_j}\text{-continuous } \left(r_j \geq 1\right) \end{array} \right. \right\} \subset \mathscr{R}^{m-1}.
\end{aligned} \tag{6.7}
$$

Definition 6.7 With $\varphi_j(\mathbf{x}^{(0,j)}, \mathbf{y}^{(0,j)}, t, \boldsymbol{\lambda}_j) = 0$, on the boundaries $\partial \mho_{\alpha_j\beta_j}$ and $\partial \underline{\mho}_{\alpha_j\beta_j}$, there is a dynamical system as

$$
\begin{aligned}
\dot{\mathbf{y}}^{(0,j)} &= \mathbf{F}^{(0,j)}\left(\mathbf{y}^{(0,j)}, t, \boldsymbol{\lambda}_j\right), \\
\dot{\mathbf{x}}^{(0,j)} &= \mathcal{F}^{(0,j)}\left(\mathbf{x}^{(0,j)}, t, \boldsymbol{\lambda}_j\right).
\end{aligned} \tag{6.8}
$$

where $\mathbf{y}^{(0,j)} = \left(y_1^{(0,j)}, y_2^{(0,j)}, \cdots, y_n^{(0,j)}\right)^{\mathrm{T}}$ and $\mathbf{x}^{(0,j)} = \left(x_1^{(0,j)}, x_2^{(0,j)}, \cdots, x_m^{(0,j)}\right)^{\mathrm{T}}$. The corresponding flow $\mathbf{y}^{(0,j)}$, (i.e., $\mathbf{y}^{(0,j)}(t) = \boldsymbol{\Phi}^{(0,j)}(\mathbf{y}^{(0,j)}(t_0), t, \boldsymbol{\lambda}_j)$) is C^{r_j+1}-continuous for time t. The corresponding flow $\mathbf{x}^{(0,j)}$ (i.e.,$\mathbf{x}^{(0,j)}(t) = \underline{\boldsymbol{\Phi}}^{(0,j)}(\mathbf{x}^{(0,j)}(t_0), t, \boldsymbol{\lambda}_j)$) is also C^{r_j+1}-continuous.

As in Luo (2009, 2012), a new vector of state variables of two dynamical systems in Eq. 6.1 is introduced as follows.

Definition 6.8 A resultant vector of state variables is

$$x = (\mathbf{x}; \mathbf{y})^{\mathrm{T}} = (x_1, x_2, \cdots, x_m; y_1, y_2, \cdots, y_n)^{\mathrm{T}} \in \mathcal{R}^{m+n}. \tag{6.9}$$

The notation $(\cdot;\cdot) \equiv (\cdot,\cdot)$ is for a combined vector of state vectors of two dynamical systems. Thus, from Luo (2009, 2012), the interaction boundary and domains are described as follows.

Definition 6.9 An interaction boundary in an $(n + m)$-dimensional phase space for the interaction of two dynamical systems in Eq. 6.1 to the interaction condition in Eq. 6.3 is defined as

$$\begin{aligned} \partial\Omega_{12} &= \overline{\Omega}_1 \cap \overline{\Omega}_2 \\ &= \left\{ x^{(0)} \left| \begin{array}{l} \varphi(x^{(0)}, t, \boldsymbol{\lambda}) \equiv \varphi(\mathbf{x}^{(0)}(t), \mathbf{y}^{(0)}(t), t, \boldsymbol{\lambda}) = 0, \\ \varphi \text{ is } C^r\text{-continuous } (r \geq 1) \end{array} \right. \right\} \\ &\subset \mathcal{R}^{n+m-1}; \end{aligned} \tag{6.10}$$

and two corresponding domains for a resultant system of two dynamical systems in Eq. 6.1 are defined as

$$\begin{aligned} \Omega_1 &= \left\{ x^{(1)} \left| \begin{array}{l} \varphi(x^{(1)}, t, \boldsymbol{\lambda}) \equiv \varphi(\mathbf{x}^{(1)}(t), \mathbf{y}^{(1)}(t), t, \boldsymbol{\lambda}) > 0, \\ \varphi \text{ is } C^r\text{-continuous } (r \geq 1) \end{array} \right. \right\} \\ &\subset \mathcal{R}^{m+n}; \\ \Omega_2 &= \left\{ x^{(2)} \left| \begin{array}{l} \varphi(x^{(2)}, t, \boldsymbol{\lambda}) \equiv \varphi(\mathbf{x}^{(2)}(t), \mathbf{y}^{(2)}(t), t, \boldsymbol{\lambda}) < 0, \\ \varphi \text{ is } C^r\text{-continuous } (r \geq 1) \end{array} \right. \right\} \\ &\subset \mathcal{R}^{m+n}. \end{aligned} \tag{6.11}$$

Definition 6.10 The resultant boundary and domains are defined by the direct product of two boundaries and domains as

$$\left.\begin{aligned}\Omega_{\alpha} &= \mho_{\alpha} \otimes \underline{\mho}_{\alpha} \quad \text{for } \alpha \in \{1,2\} \\ \partial\Omega_{\alpha\beta} &= \partial\mho_{\alpha\beta} \otimes \partial\underline{\mho}_{\alpha\beta} \quad \text{for } \alpha,\beta \in \{1,2\}\end{aligned}\right\} \tag{6.12}$$

On the two domains, a resultant system of two dynamical systems is discontinuous to the interaction boundary, defined by

$$\dot{x}^{(\alpha)} = \mathbb{F}^{(\alpha)}\left(x^{(\alpha)}, t, \boldsymbol{\pi}^{(\alpha)}\right) \quad \text{in } \Omega_{\alpha} \; (\alpha = 1, 2) \tag{6.13}$$

where

$$\begin{gathered}\mathbb{F}^{(\alpha)} = \left(\boldsymbol{\mathcal{F}}^{(\alpha)}; \mathbf{F}^{(\alpha)}\right)^{\mathrm{T}} \\ = \left(\mathcal{F}_{1}^{(\alpha)}, \mathcal{F}_{2}^{(\alpha)}, \cdots, \mathcal{F}_{m}^{(\alpha)}; F_{1}^{(\alpha)}, F_{2}^{(\alpha)} \cdots, F_{n}^{(\alpha)}\right)^{\mathrm{T}} \\ \boldsymbol{\pi}^{(\alpha)} = (\mathbf{q}_{\alpha}, \mathbf{p}_{\alpha})^{\mathrm{T}}.\end{gathered} \tag{6.14}$$

Suppose there is a vector field $\mathbb{F}^{(0)}\left(x^{(0)}, t, \boldsymbol{\lambda}\right)$ on the interaction boundary with $\varphi\left(x^{(0)}, t, \boldsymbol{\lambda}\right) = 0$, and the corresponding dynamical system on such a boundary is expressed by

$$\dot{x}^{(0)} = \mathbb{F}^{(0)}\left(x^{(0)}, t, \boldsymbol{\lambda}\right) \quad \text{on } \partial\Omega_{12}. \tag{6.15}$$

From Eq. 6.4, there are many conditions for interactions of two dynamical systems. Suppose only the *j*th interaction boundary occurs for two system for time *t*. From Luo (2009, 2012), the above definitions can be extended as follows.

Definition 6.11 The *j*th interaction boundary in an $(n + m)$-dimensional phase space for the interaction of two dynamical systems in Eq. 6.1, relative to the *j*th constraint of the interaction conditions in Eq. 6.3, is defined as

$$\begin{gathered}\partial\Omega_{(\alpha_j\beta_j j)} = \overline{\Omega}_{(\alpha_j,j)} \cap \overline{\Omega}_{(\beta_j,j)} \\ = \left\{ x^{(0,j)} \left| \begin{aligned} &\varphi_j\left(x^{(0,j)}, t, \boldsymbol{\lambda}_j\right) \\ &\equiv \varphi_j\left(\mathbf{x}^{(0,j)}(t), \mathbf{y}^{(0,j)}(t), t, \boldsymbol{\lambda}_j\right) = 0, \\ &\varphi_j \text{ is } C^{r_j}\text{-continuous } \left(r_j \geq 1\right) \end{aligned} \right. \right\} \\ \subset \mathcal{R}^{m+n-1};\end{gathered} \tag{6.16}$$

and two domains pertaining to the *j*th boundary for a resultant system of two dynamical systems in Eq. 6.1 are defined as

$$\left.\begin{aligned}\Omega_{(1,j)} &= \left\{ x^{(1,j)} \left| \begin{aligned} &\varphi_j\left(x^{(1,j)}, t, \boldsymbol{\lambda}_j\right) \\ &\equiv \varphi_j\left(\mathbf{x}^{(1,j)}(t), \mathbf{y}^{(1,j)}(t), t, \boldsymbol{\lambda}_j\right) > 0, \\ &\varphi_j \text{ is } C^{r_j}\text{-continuous } \left(r_j \geq 1\right) \end{aligned} \right. \right\} \subset \mathcal{R}^{m+n}; \\ \Omega_{(2,j)} &= \left\{ x^{(2,j)} \left| \begin{aligned} &\varphi_j\left(x^{(2,j)}, t, \boldsymbol{\lambda}_j\right) \\ &\equiv \varphi_j\left(\mathbf{x}^{(2,j)}(t), \mathbf{y}^{(2,j)}(t), t, \boldsymbol{\lambda}_j\right) < 0, \\ &\varphi_j \text{ is } C^{r_j}\text{-continuous } \left(r_j \geq 1\right) \end{aligned} \right. \right\} \subset \mathcal{R}^{m+n}. \end{aligned}\right. \tag{6.17}$$

Definition 6.12 Such a boundary and domains for the jth interaction condition are defined expressed by the direct product of two corresponding individual boundaries or domains

$$\left.\begin{aligned} \Omega_{(\alpha_j,j)} &= \mho_{(\alpha_j,j)} \otimes \underline{\mho}_{(\alpha_j,j)} \quad \text{for } \alpha_j \in \{1,2\}, \\ \partial\Omega_{(\alpha_j\beta_j,j)} &= \partial\mho_{(\alpha_j\beta_j,j)} \otimes \partial\underline{\mho}_{(\alpha_j\beta_j,j)} \quad \text{for } \alpha_j, \beta_j \in \{1,2\} \end{aligned}\right\} \tag{6.18}$$

On the two domains relative to the jth interaction boundary, a discontinuous resultant system of two dynamical systems in Eq. 6.1 with the jth interaction in Eq. 6.4 is defined by

$$\dot{x}^{(\alpha_j,j)} = \mathbb{F}^{(\alpha_j,j)}\left(x^{(\alpha_j,j)}, t, \boldsymbol{\pi}_j^{(\alpha_j)}\right) \quad \text{in } \Omega_{(\alpha_j,j)} \tag{6.19}$$

where $\mathbb{F}^{(\alpha_j,j)} = \left(\boldsymbol{\mathcal{F}}^{(\alpha_j,j)}; \mathbf{F}^{(\alpha_j,j)}\right)^{\mathrm{T}}$ and $\boldsymbol{\pi}_j^{(\alpha_j)} = \left(\mathbf{q}_j^{(\alpha_j)}, \mathbf{p}_j^{(\alpha_j)}\right)^{\mathrm{T}}$. Suppose there is a vector field $\mathbb{F}^{(0,j)}\left(x^{(0,j)}, t, \boldsymbol{\lambda}_j\right)$ on the jth interaction boundary with $\varphi_j\left(x^{(0,j)}, t, \boldsymbol{\lambda}_j\right) = 0$, and the corresponding dynamical system on the jth interaction boundary is expressed by

$$\dot{x}^{(0,j)} = \mathbb{F}^{(0,j)}\left(x^{(0,j)}, t, \boldsymbol{\lambda}_j\right) \quad \text{on } \partial\Omega_{(12,j)}. \tag{6.20}$$

6.2 An Interaction Theory

As in Luo (2009, 2012), resultant state variables in phase space are introduced.

Definition 6.13 New variables in domain $\Omega_{(\alpha_j,j)}$ are defined as

$$z^{(\alpha_j,j)} = \varphi_j\left(\mathbf{x}^{(\alpha_j,j)}(t), \mathbf{y}^{(\alpha_j,j)}(t), t, \boldsymbol{\lambda}_j\right) \text{ for } j \in \mathcal{L}.. \tag{6.21}$$

From the above new variables, the resultant dynamical system can be defined as in Luo (2009, 2012).

Definition 6.14 For $\mathbf{g}^{(\alpha_j,j)} = \left(g_1^{(\alpha_j,j)}, g_2^{(\alpha_j,j)}\right)^{\mathrm{T}}$, the resultant dynamical system is defined as

$$\left.\begin{aligned} &\dot{\mathbf{z}}^{(\alpha_j,j)} = \mathbf{g}^{(\alpha_j,j)}\left(\mathbf{z}^{(\alpha_j,j)}, t\right) \text{ for } j \in \mathcal{L};\\ &\dot{\mathbf{x}}^{(\alpha_j,j)} = \boldsymbol{\mathcal{F}}^{(\alpha_j,j)}\left(\mathbf{x}^{(\alpha_j,j)}, t, \mathbf{q}^{(\alpha_j,j)}\right) \in \mathcal{R}^m,\\ &\dot{\mathbf{y}}^{(\alpha_j,j)} = \mathbf{F}^{(\alpha_j,j)}\left(\mathbf{y}^{(\alpha_j,j)}, t, \mathbf{p}^{(\alpha_j,j)}\right) \in \mathcal{R}^n. \end{aligned}\right\} \tag{6.22}$$

For a better understanding of such a discontinuous dynamical system, Luo (2009, 2012) defined the resultant boundary and domains as follows.

Definition 6.15 The boundary and domains in phase space are defined as

$$\begin{aligned} \partial\Xi_{(\alpha_j\beta_j,j)} &= \overline{\Xi}_{(\alpha_j,j)} \cap \overline{\Xi}_{(\beta_j,j)}\\ &= \left\{\left(z^{(0,j)}, \dot{z}^{(0,j)}\right) \middle| \psi_j\left(z^{(0,j)}, \dot{z}^{(0,j)}\right) = z^{(0,j)} = 0\right\}\\ &\subset \mathcal{R}; \end{aligned} \tag{6.23}$$

and

$$\begin{aligned} \Xi_{(1,j)} &= \left\{\left(z^{(1,j)}, \dot{z}^{(1,j)}\right) \middle| z^{(1,j)} > 0\right\} \subset \mathcal{R}^2;\\ \Xi_{(2,j)} &= \left\{\left(z^{(1,j)}, \dot{z}^{(1,j)}\right) \middle| z^{(2,j)} < 0\right\} \subset \mathcal{R}^2. \end{aligned} \tag{6.24}$$

The interaction boundary is determined by

$$\begin{aligned} &\left.\begin{aligned} &z^{(0,j)} = \varphi_j\left(\mathbf{x}^{(0,j)}, \mathbf{y}^{(0,j)}, t, \boldsymbol{\lambda}_j\right) = 0,\\ &\dot{z}^{(0,j)} = D\varphi_j\left(\mathbf{x}^{(0,j)}, \mathbf{y}^{(0,j)}, t, \boldsymbol{\lambda}_j\right) = 0 \end{aligned}\right\} \text{ for } j \in \mathcal{L};;\\ &\dot{\mathbf{x}}^{(0,j)} = \boldsymbol{\mathcal{F}}^{(0,j)}\left(\mathbf{x}^{(0,j)}, t, \boldsymbol{\lambda}_j\right) \in \mathcal{R}^n,\\ &\dot{\mathbf{y}}^{(0,j)} = \mathbf{F}^{(0,j)}\left(\mathbf{y}^{(0,j)}, t, \boldsymbol{\lambda}_j\right) \in \mathcal{R}^m. \end{aligned} \tag{6.25}$$

From the above description, Luo (2009, 2012) introduced a new dynamical system for system interaction as follows.

Definition 6.15 The dynamical systems in phase space $(z, \dot{z})$ are defined as follows.

$$\begin{aligned}
&\dot{\mathbf{z}}^{(\Lambda_j,j)} = \mathbf{g}^{(\Lambda_j,j)}\left(\mathbf{z}^{(\Lambda_j,j)}, t\right) \text{ for } j \in \mathcal{L}, \Lambda_j = 0, \alpha_j \\
&\dot{\mathbf{x}}^{(\Lambda_j,j)} = \boldsymbol{\mathcal{F}}^{(\Lambda_j,j)}\left(\mathbf{x}^{(\Lambda_j,j)}, t, \boldsymbol{\lambda}_j\right) \in \mathcal{R}^n, \\
&\dot{\mathbf{y}}^{(\Lambda_j,j)} = \mathbf{F}^{(\Lambda_j,j)}\left(\mathbf{y}^{(\Lambda_j,j)}, t, \boldsymbol{\lambda}_j\right) \in \mathcal{R}^m.
\end{aligned} \tag{6.26}$$

where

$$\left.\begin{aligned}
&\mathbf{g}^{(\alpha_j,j)}\left(\mathbf{z}^{(\alpha_j,j)}, t\right) = \left(g_1^{(\alpha_j,j)}\left(\mathbf{z}^{(\alpha_j,j)}, t\right), g_2^{(\alpha_j,j)}\left(\mathbf{z}^{(\alpha_j,j)}, t\right)\right)^{\mathrm{T}} \\
&\qquad \text{in } \Xi_{\alpha_j}\left(\alpha_j \in \{1,2\}\right); \\
&\mathbf{g}^{(0,j)}\left(\mathbf{z}^{(\alpha_j,j)}, t\right) \in \left[\mathbf{g}^{(\alpha_j,j)}\left(\mathbf{z}^{(\alpha_j,j)}, t\right), \mathbf{g}^{(\beta_j,j)}\left(\mathbf{z}^{(\beta_j,j)}, t\right)\right] \\
&\qquad \text{on } \partial\Xi_{(\alpha_j\beta_j,j)} \text{ for non-stick}, \\
&\mathbf{g}^{(0,j)}\left(\mathbf{z}^{(\alpha_j,j)}, t\right) = (0,0)^T \text{on } \partial\Xi_{(\alpha_j\beta_j,j)} \text{ for stick.}
\end{aligned}\right\} \tag{6.27}$$

The normal vector of $\partial\Xi_{(\alpha_j\beta_j,j)}$ is computed from Eq. 6.23, i.e.,

$$\mathbf{n}_{\partial\Xi_{(\alpha_j\beta_j,j)}} = (1,0)^{\mathrm{T}} \text{ and } D\mathbf{n}_{\partial\Xi_{(\alpha_j\beta_j,j)}} = (0,0)^{\mathrm{T}}, \tag{6.28}$$

where $D(\cdot) = D(\cdot)/Dt$. From Luo (2009, 2012), the corresponding two G-functions are obtained by

$$\begin{aligned}
G_{\partial\Xi_{(12,j)}}^{(0,\alpha_j)}\left(\mathbf{z}^{(\alpha_j,j)}, t\right) &= \mathbf{n}_{\partial\Xi_{(12,j)}} \cdot \mathbf{g}^{(\alpha_j,j)}\left(\mathbf{z}^{(\alpha_j,j)}, t\right) \\
&= g_1^{(\alpha_j,j)}\left(\mathbf{z}^{(\alpha_j,j)}, t\right), \\
G_{\partial\Xi_{(12,j)}}^{(1,\alpha_j)}\left(\mathbf{z}^{(\alpha_j,j)}, t\right) &= \mathbf{n}_{\partial\Xi_{(12,j)}} \cdot D\mathbf{g}^{(\alpha_j,j)}\left(\mathbf{z}^{(\alpha_j,j)}, t\right) \\
&= g_2^{(\alpha_j,j)}\left(\mathbf{z}^{(\alpha_j,j)}, t\right).
\end{aligned} \tag{6.29}$$

From the above description and application of the discontinuous dynamical systems, the analytical conditions for system interactions are given in Luo (2009, 2012).

(i) The sufficient and necessary conditions for a passable flow at $\left(\mathbf{z}_m^{(0,j)}, t_m\right)$ and $\left(\mathbf{z}_m^{(\alpha_j,j)}, t_m\right)$ for the boundary $\partial\Xi_{(12,j)}$ are given from Luo (2009, 2012) as follows:

$$\left.\begin{aligned} &G_{\partial\Xi_{(12,j)}}^{(0,1)}\left(\mathbf{z}_m^{(1,j)}, t_{m-}\right) = g_1^{(1,j)}\left(\mathbf{z}_m^{(1,j)}, t_{m-}\right) < 0, \\ &G_{\partial\Xi_{(12,j)}}^{(0,2)}\left(\mathbf{z}_m^{(2,j)}, t_{m+}\right) = g_1^{(2,j)}\left(\mathbf{z}_m^{(2,j)}, t_{m+}\right) < 0 \end{aligned}\right\} \text{from } \Xi_{(1,j)} \to \Xi_{(2,j)}$$
$$\left.\begin{aligned} &G_{\partial\Xi_{(12,j)}}^{(0,1)}\left(\mathbf{z}_m^{(1,j)}, t_{m+}\right) = g_1^{(1,j)}\left(\mathbf{z}_m^{(1,j)}, t_{m+}\right) > 0, \\ &G_{\partial\Xi_{(12,j)}}^{(0,2)}\left(\mathbf{z}_m^{(2,j)}, t_{m-}\right) = g_1^{(2,j)}\left(\mathbf{z}_m^{(2,j)}, t_{m-}\right) > 0 \end{aligned}\right\} \text{from } \Xi_{(2,j)} \to \Xi_{(1,j)} \tag{6.30}$$

where

$$\begin{aligned} g_1^{(\alpha_j,j)}\left(\mathbf{z}_m^{(\alpha_j,j)}, t_{m\pm}\right) &= \underline{\mathbf{n}}_{\varphi_j} \cdot \mathcal{F}^{(\alpha_j,j)}\left(\mathbf{x}_m^{(\alpha_j,j)}, t_{m\pm}, \mathbf{q}^{(\alpha_j,j)}\right) \\ &+ \mathbf{n}_{\varphi_j} \cdot \mathbf{F}^{(\alpha_j,j)}\left(\mathbf{y}_m^{(\alpha_j,j)}, t_{m\pm}, \mathbf{p}^{(\alpha_j,j)}\right) + \frac{\partial \varphi_j}{\partial t}. \end{aligned} \tag{6.31}$$

As in Luo (2009, 2012), the foregoing condition gives the sufficient and necessary conditions for two systems to interact under the jth interaction condition and to switch the current states through such an interaction condition. Such a flow to the boundary is called an *instantaneous interaction* between two systems.

(ii) The sufficient and necessary conditions for a stick flow (or sink flow) on the boundary $\partial\Xi_{(\alpha_j\beta_j,j)}$ are given from Luo (2009, 2012) as follows:

$$\left.\begin{aligned} &G_{\partial\Xi_{(12,j)}}^{(0,1)}\left(\mathbf{z}_m^{(1,j)}, t_{m-}\right) = g_1^{(1,j)}\left(\mathbf{z}_m^{(1,j)}, t_{m-}\right) < 0, \\ &G_{\partial\Xi_{(12,j)}}^{(0,2)}\left(\mathbf{z}_m^{(2,j)}, t_{m-}\right) = g_1^{(2,j)}\left(\mathbf{z}_m^{(2,j)}, t_{m-}\right) > 0 \end{aligned}\right\} \text{ on } \partial\Xi_{(12,j)}. \tag{6.32}$$

From the foregoing condition, the two systems will stick together under the jth interaction condition, which is called a *stick interaction*.

(iii) The sufficient and necessary conditions for a source flow on the boundary $\partial\Xi_{(12,j)}$ are given from Luo (2009, 2012) as follows:

$$\left.\begin{aligned} G_{\partial \Xi_{(12,j)}}^{(0,1)}\left(\mathbf{z}_m^{(1,j)}, t_{m+}\right) = g_1^{(1,j)}\left(\mathbf{z}_m^{(1,j)}, t_{m+}\right) > 0, \\ G_{\partial \Xi_{(12,j)}}^{(0,2)}\left(\mathbf{z}_m^{(2,j)}, t_{m+}\right) = g_1^{(2,j)}\left(\mathbf{z}_m^{(2,j)}, t_{m+}\right) < 0 \end{aligned}\right\} \text{ on } \partial\Xi_{(12,j)}. \tag{6.33}$$

From Luo (2009, 2012), based on the above conditions, the two dynamical systems will not interact at $(\mathbf{z}_m^{(0,j)}, t_m)$ for the boundary $\partial\Xi_{(\alpha_j\beta_j j)}$ relative to the jth interaction condition. The phenomenon is called the *source* interaction (or de-interaction).

The appearance and disappearance of three interaction states of the two dynamical systems to the jth interaction condition in Eq. 6.3 can be determined from Luo (2009, 2012) for the switching bifurcation of three states of interactions between the two dynamical systems.

(iv) For appearance of the stick interaction from the instantaneous interaction, the sufficient and necessary conditions are given from Luo (2009, 2012) as follows:

$$\begin{aligned} &(-1)^{\alpha_j} G_{\partial \Xi_{(\alpha_j\beta_j j)}}^{(0,\alpha_j)}\left(\mathbf{z}_m^{(\alpha_j,j)}, t_{m-}\right) = (-1)^{\alpha_j} g_1^{(\alpha_j,j)}\left(\mathbf{z}_m^{(\alpha_j,j)}, t_{m-}\right) > 0, \\ &G_{\partial \Xi_{(\alpha_j\beta_j j)}}^{(0,\beta_j)}\left(\mathbf{z}_m^{(\beta_j,j)}, t_{m\pm}\right) = g_1^{(\beta_j,j)}\left(\mathbf{z}_m^{(\beta_j,j)}, t_{m\pm}\right) = 0, \\ &(-1)^{\beta_j} G_{\partial \Xi_{(\alpha_j\beta_j j)}}^{(1,\beta_j)}\left(\mathbf{z}_m^{(\beta_j,j)}, t_{m\pm}\right) = (-1)^{\beta_j} g_2^{(\beta_j,j)}\left(\mathbf{z}_m^{(\beta_j,j)}, t_{m\pm}\right) < 0. \end{aligned} \tag{6.34}$$

(v) For the vanishing of the stick interaction to form the instantaneous interaction on the jth interaction boundary, the sufficient and necessary conditions are given from Luo (2009, 2012) as follows:

$$\begin{aligned} &(-1)^{\alpha_j} G_{\partial \Xi_{(\alpha_j\beta_j j)}}^{(0,\alpha_j)}\left(\mathbf{z}_m^{(\alpha_j,j)}, t_{m-}\right) = (-1)^{\alpha_j} g_1^{(\alpha_j,j)}\left(\mathbf{z}_m^{(\alpha_j,j)}, t_{m-}\right) > 0, \\ &G_{\partial \Xi_{(\alpha_j\beta_j j)}}^{(0,\beta_j)}\left(\mathbf{z}_m^{(\beta_j,j)}, t_{m\mp}\right) = g_1^{(\beta_j,j)}\left(\mathbf{z}_m^{(\beta_j,j)}, t_{m\mp}\right) = 0, \\ &(-1)^{\beta_j} G_{\partial \Xi_{(\alpha_j\beta_j j)}}^{(1,\beta_j)}\left(\mathbf{z}_m^{(\beta_j,j)}, t_{m\mp}\right) = (-1)^{\beta_j} g_2^{(\beta_j,j)}\left(\mathbf{z}_m^{(\beta_j,j)}, t_{m\mp}\right) < 0. \end{aligned} \tag{6.35}$$

From Luo (2009, 2012), the sufficient and necessary conditions for appearance and vanishing of the *source* interaction, pertaining to the instantaneous interaction, are obtained.

(vi) The appearance of source interaction is given from Luo (2009, 2012) through the following conditions.

$$
\begin{aligned}
&(-1)^{\alpha_j} G_{\partial\Xi_{(\alpha_i\beta_j,j)}}^{(0,\alpha_j)}\left(\mathbf{z}_m^{(\alpha_j,j)},t_{m+}\right) = (-1)^{\alpha_j} g_1^{(\alpha_j,j)}\left(\mathbf{z}_m^{(\alpha_j,j)},t_{m+}\right) < 0,\\
&G_{\partial\Xi_{(\alpha_i\beta_j,j)}}^{(0,\beta_j)}\left(\mathbf{z}_m^{(\beta_j,j)},t_{m\mp}\right) = g_1^{(\beta_j,j)}\left(\mathbf{z}_m^{(\beta_j,j)},t_{m\mp}\right) = 0,\\
&(-1)^{\beta_j} G_{\partial\Xi_{(\alpha_i\beta_j,j)}}^{(1,\beta_j)}\left(\mathbf{z}_m^{(\beta_j,j)},t_{m\mp}\right) = (-1)^{\beta_j} g_2^{(\beta_j,j)}\left(\mathbf{z}_m^{(\beta_j,j)},t_{m\mp}\right) < 0.
\end{aligned}
\tag{6.36}
$$

(vii) The vanishing of the source interaction is given from Luo (2009, 2012) through the following conditions.

$$
\begin{aligned}
&(-1)^{\alpha_j} G_{\partial\Xi_{(\alpha_i\beta_j,j)}}^{(0,\alpha_j)}\left(\mathbf{z}_m^{(\alpha_j,j)},t_{m+}\right) = (-1)^{\alpha_j} g_1^{(\alpha_j,j)}\left(\mathbf{z}_m^{(\alpha_j,j)},t_{m+}\right) < 0,\\
&G_{\partial\Xi_{(\alpha_i\beta_j,j)}}^{(0,\beta_j)}\left(\mathbf{z}_m^{(\beta_j,j)},t_{m\pm}\right) = g_1^{(\beta_j,j)}\left(\mathbf{z}_m^{(\beta_j,j)},t_{m\pm}\right) = 0,\\
&(-1)^{\beta_j} G_{\partial\Xi_{(\alpha_i\beta_j,j)}}^{(1,\beta_j)}\left(\mathbf{z}_m^{(\beta_j,j)},t_{m\pm}\right) = (-1)^{\beta_j} g_2^{(\beta_j,j)}\left(\mathbf{z}_m^{(\beta_j,j)},t_{m\pm}\right) < 0.
\end{aligned}
\tag{6.37}
$$

(viii) The sufficient and necessary conditions for the switching between the stick and source interactions on the jth interaction boundary are given from Luo (2009, 2012) through the following conditions.

$$
\left.
\begin{aligned}
&G_{\partial\Xi_{(\alpha_i\beta_j,j)}}^{(0,\alpha_j)}\left(\mathbf{z}_m^{(\alpha_j,j)},t_{m\mp}\right) = g_1^{(\alpha_j,j)}\left(\mathbf{z}_m^{(\alpha_j,j)},t_{m\mp}\right) = 0\\
&(-1)^{\alpha_j} G_{\partial\Xi_{(\alpha_i\beta_j,j)}}^{(1,\alpha_j)}\left(\mathbf{z}_m^{(\alpha_j,j)},t_{m\mp}\right) = (-1)^{\alpha_j} g_2^{(\alpha_j,j)}\left(\mathbf{z}_m^{(\alpha_j,j)},t_{m\mp}\right) < 0;\\
&G_{\partial\Xi_{(\alpha_i\beta_j,j)}}^{(0,\beta_j)}\left(\mathbf{z}_m^{(\beta_j,j)},t_{m\mp}\right) = g_1^{(\beta_j,j)}\left(\mathbf{z}_m^{(\beta_j,j)},t_{m\mp}\right) = 0,\\
&(-1)^{\beta_j} G_{\partial\Xi_{(\alpha_i\beta_j,j)}}^{(1,\beta_j)}\left(\mathbf{z}_m^{(\beta_j,j)},t_{m\mp}\right) = (-1)^{\beta_j} g_2^{(\beta_j,j)}\left(\mathbf{z}_m^{(\beta_j,j)},t_{m\mp}\right) < 0.
\end{aligned}
\right\}
\tag{6.38}
$$

(ix) The sufficient and necessary conditions for the switching between two instantaneous interaction states on the jth interaction boundary for $\alpha_j \neq \beta_j$ are given from Luo (2009, 2012) through the following conditions.

$$\left.\begin{aligned} &G^{(0,\alpha_j)}_{\partial\Xi_{(\alpha_i\beta_j,j)}}\left(\mathbf{z}_m^{(\alpha_j,j)},t_{m\mp}\right)=g_1^{(\alpha_j,j)}\left(\mathbf{z}_m^{(\alpha_j,j)},t_{m\mp}\right)=0 \text{ for } \alpha_j\in\{1,2\};\\ &(-1)^{\alpha_j}G^{(1,\alpha_j)}_{\partial\Xi_{(\alpha_i\beta_j,j)}}\left(\mathbf{z}_m^{(\alpha_j,j)},t_{m\mp}\right)=(-1)^{\alpha_j}g_2^{(\alpha_j,j)}\left(\mathbf{z}_m^{(\alpha_j,j)},t_{m\mp}\right)<0;\\ &G^{(0,\beta_j)}_{\partial\Xi_{(\alpha_i\beta_j,j)}}\left(\mathbf{z}_m^{(\beta_j,j)},t_{m\pm}\right)=g_1^{(\beta_j,j)}\left(\mathbf{z}_m^{(\beta_j,j)},t_{m\pm}\right)=0 \text{ for } \beta_j\in\{1,2\};\\ &(-1)^{\beta_j}G^{(1,\beta_j)}_{\partial\Xi_{(\alpha_i\beta_j,j)}}\left(\mathbf{z}_m^{(\beta_j,j)},t_{m\pm}\right)=(-1)^{\beta_j}g_2^{(\beta_j,j)}\left(\mathbf{z}_m^{(\beta_j,j)},t_{m\pm}\right)<0. \end{aligned}\right\} \tag{6.39}$$

From Luo (2009, 2012), a flow tangential to the boundary $\partial\Xi_{(\alpha_j\beta_j,j)}$ is another instantaneous interaction (or tangential interaction).

(x) The sufficient and necessary conditions for the tangential interactions are given from Luo (2009, 2012) through the following conditions.

$$\left.\begin{aligned} &G^{(0,\alpha_j)}_{\partial\Xi_{(\alpha_i\beta_j,j)}}\left(\mathbf{z}_m^{(\alpha_j,j)},t_{m\pm}\right)=g_1^{(\alpha_j,j)}\left(\mathbf{z}_m^{(\alpha_j,j)},t_{m\pm}\right)=0 \text{ for } \alpha_j\in\{1,2\};\\ &(-1)^{\alpha_j}G^{(1,\alpha_j)}_{\partial\Xi_{(\alpha_i\beta_j,j)}}\left(\mathbf{z}_m^{(\alpha_j,j)},t_{m\pm}\right)=(-1)^{\alpha_j}g_2^{(\alpha_j,j)}\left(\mathbf{z}_m^{(\alpha_j,j)},t_{m\pm}\right)<0. \end{aligned}\right\} \tag{6.40}$$

The interaction theory with higher-order singularity can be found from Luo (2009, 2012), which will not be presented herein.

6.3 Edge Interaction Theory

From Luo (2009, 2012), suppose there are s-linear-independent interaction conditions among l-interaction conditions in Eq. 6.3, such s-interaction conditions form an edge (or a singular edge). The following definition is given as in Luo(2009, 2012).

Definition 6.16 On the edge, the conditions presented in this section should be checked for all s-interaction boundaries via

$$z^{(\alpha_j,j)}=\varphi_j\left(\mathbf{x}^{(\alpha_j,j)},\mathbf{y}^{(\alpha_j,j)},t,\boldsymbol{\lambda}_j\right) \text{ for all } j\in\mathcal{L}.. \tag{6.41}$$

On the boundary $\partial\Omega_{(\alpha_j\beta_j,j)}$, one obtains

$$z^{(0,j)}=\varphi_j\left(\mathbf{x}^{(0,j)},\mathbf{y}^{(0,j)},t,\boldsymbol{\lambda}_j\right)=0 \text{ for all } j\in\mathcal{L}. \tag{6.42}$$

As in Luo (2009, 2012), without loss of generality, suppose the l linearly independent interaction constraints form an edge and there are three subsets $\mathcal{L}_i \subseteq \mathcal{L}(i=1,$ $2, 3)$ with l_i-interaction conditions among l-conditions and $l_1 + l_2 + l_3 = l$. For $j \in \mathcal{L}_i$, the conditions for l_1-stick interaction,l_2-noninteraction, and l_3-instantaneous interaction are given from Luo (2009, 2012) as follows.

Theorem 6.1 *Consider two dynamical systems in Eq. 6.1 with constraints in Eq. 6.3. Let* $\mathcal{L}_i = \varnothing \cup \left\{k_1^{(i)}, k_2^{(i)}, \cdots, k_{l_i}^{(i)}\right\}$ $(i = 1, 2, 3)$ *with* $k_\kappa^{(i)} \in \mathcal{L}(\kappa = 1, 2, \cdots l_i)$ *and* $l_1 + l_2 + l_3 = l$. *For* $j \in \mathcal{L}_i \subseteq \mathcal{L}$ *and* $\mathcal{L} = \cup_{i=1}^3 \mathcal{L}_i$, $x_m^{(\alpha_j,j)} = \left(\mathbf{x}^{(\alpha_j,j)}, \mathbf{y}^{(\alpha_j,j)}\right)^{\mathrm{T}} \in \Omega_{(\alpha_j,j)}$ $(\alpha_j \in \mathcal{I}$ *with* $\mathcal{I} = \{1,2\})$ *and* $x_m^{(0,j)} = \left(\mathbf{x}_m^{(0,j)}, \mathbf{y}_m^{(0,j)}\right)^{\mathrm{T}} \in \partial\Omega_{(12,j)}$ *at time* t_m, $x_m^{(\alpha_j,j)} = x_m^{(0,j)}$. *For any small* $\varepsilon > 0$, *there is a time interval* $[t_{m-\varepsilon}, t_m)$ *or* $(t_m, t_{m+\varepsilon}]$. *At* $x^{(\alpha_j,j)} \in \Omega_{(\alpha_j,j)}^{\pm\varepsilon}$ *for time* $t \in [t_{m-\varepsilon}, t_m)$ *or* $(t_m, t_{m+\varepsilon}]$, $z^{(\alpha_j,j)}(t) = \varphi^{(\alpha_j,j)}\left(x^{(\alpha_j,j)}(t), t, \boldsymbol{\lambda}_j\right)$ *is* $C^{r_{\alpha_{j1}}}$ *continuous and* $\left| D^{(r_{\alpha_j}+1)} z^{(\alpha_j,j)}(t) \right| < \infty$ $(r_{\alpha_j} \geq 3)$. *For* $x^{(\alpha_j,j)} \in \Omega_{(\alpha_j,j)}$ *and* $x^{(0,j)} \in \partial\Omega_{(12,j)}$. $\mathbb{F}^{(\alpha_j,j)}\left(x^{(\alpha_j,j)}, t, \boldsymbol{\pi}^{(\alpha_j,j)}\right) \neq \mathbb{F}^{(0,j)}\left(x^{(0,j)}, t, \boldsymbol{\lambda}_j\right)$ *at* $x^{(\alpha_j,j)} = x^{(0,j)}$. *The two dynamical systems in Eq. 6.1 to the corner points of l-interaction conditions in Eq. 6.3 are of the* (l_1, l_2, l_3)*-stick interaction, noninteraction, and instantaneous interaction for time* t_m *if and only if*

(i) *For all* $j \in \mathcal{L}_1$ *with* $\alpha_j = 1, 2$ *at time* t_m

$$\left.\begin{array}{l} x_{m-}^{(\alpha_j,j)} = x_m^{(0,j)} \left(\text{or } z_{m-}^{(\alpha_j,j)} = z_m^{(0,j)}\right), \\ (-1)^{\alpha_j} G_{\partial\Xi_{(\alpha_j\beta_j,j)}}^{(0,j)}\left(\mathbf{z}_m^{(\alpha_j,j)}, t_{m-}\right) = (-1)^{\alpha_j} g_1^{(\alpha_j,j)}\left(\mathbf{z}_m^{(\alpha_j,j)}, t_{m-}\right) > 0. \end{array}\right\} \tag{6.43}$$

(ii) *For all* $j \in \mathcal{L}_2$ *with* $\alpha_j = 1, 2$ *at time* t_m

$$\left.\begin{array}{l} x_{m+}^{(\alpha_j,j)} = x_m^{(0,j)} \left(\text{or } z_{m+}^{(\alpha_j,j)} = z_m^{(0,j)}\right), \\ (-1)^{\alpha_j} G_{\partial\Xi_{(\alpha_j\beta_j,j)}}^{(0,j)}\left(\mathbf{z}_m^{(\alpha_j,j)}, t_{m+}\right) = (-1)^{\alpha_j} g_1^{(\alpha_j,j)}\left(\mathbf{z}_m^{(\alpha_j,j)}, t_{m+}\right) < 0. \end{array}\right\} \tag{6.44}$$

(iii) *For all* $j \in \mathcal{L}_3$ *with* $\alpha_j, \beta_j = 1, 2$ *and* $\alpha_j \neq \beta_j$ *at time* t_m

$$
\left.\begin{aligned}
&x_{m-}^{(\alpha_j,j)}=x_m^{(0,j)}\left(\text{or } z_{m-}^{(\alpha_j,j)}=z_m^{(0,j)}\right),\\
&(-1)^{\alpha_j}G_{\partial\Xi_{(\alpha_j\beta_j,j)}}^{(0,j)}\left(\mathbf{z}_m^{(\alpha_j,j)},t_{m-}\right)=(-1)^{\alpha_j}g_1^{(\alpha_j,j)}\left(\mathbf{z}_m^{(\alpha_j,j)},t_{m-}\right)>0;\\
&x_{m+}^{(\beta_j,j)}=x_m^{(0,j)}\left(\text{or } z_{m+}^{(\beta_j,j)}=z_m^{(0,j)}\right),\\
&(-1)^{\beta_j}G_{\partial\Xi_{(\alpha_j\beta_j,j)}}^{(0,j)}\left(\mathbf{z}_m^{(\beta_j,j)},t_{m+}\right)=(-1)^{\beta_j}g_1^{(\beta_j,j)}\left(\mathbf{z}_m^{(\beta_j,j)},t_{m+}\right)<0.
\end{aligned}\right\}\tag{6.45}
$$

(iv) *The switching bifurcation conditions for one of three cases (i–iii) for $j \in \mathcal{L}_i$ $(i = 1, 2, 3)$ with time $t = t_m$ with $\alpha_j \in \{1,2\}$,*

$$
\left.\begin{aligned}
&x_{m\pm}^{(\alpha_j,j)}=x_m^{(0,j)}\left(\text{or } z_{m\pm}^{(\alpha_j,j)}=z_m^{(0,j)}\right),\\
&G_{\partial\Xi_{(\alpha_j\beta_j,j)}}^{(0,j)}\left(\mathbf{z}_m^{(\alpha_j,j)},t_{m\pm}\right)=g_1^{(\alpha_j,j)}\left(\mathbf{z}_m^{(\alpha_j,j)},t_{m\pm}\right)=0;\\
&(-1)^{\alpha_j}G_{\partial\Xi_{(\alpha_j\beta_j,j)}}^{(1,j)}\left(\mathbf{z}_m^{(\alpha_j,j)},t_{m\pm}\right)=(-1)^{\alpha_j}g_2^{(\alpha_j,j)}\left(\mathbf{z}_m^{(\alpha_j,j)},t_{m\pm}\right)<0.
\end{aligned}\right\}\tag{6.46}
$$

and for $\beta_j = \{1,2\}$ but $\beta_j \neq \alpha_j$

$$
\begin{aligned}
&x_{m\pm}^{(\beta_j,j)}=x_m^{(0,j)}\left(\text{or } z_{m\pm}^{(\beta_j,j)}=z_m^{(0,j)}\right),\\
&G_{\partial\Xi_{(\alpha_j\beta_j,j)}}^{(0,j)}\left(\mathbf{z}_m^{(\beta_j,j)},t_{m\pm}\right)=g_1^{(\beta_j,j)}\left(\mathbf{z}_m^{(\beta_j,j)},t_{m\pm}\right)\neq 0,
\end{aligned}\tag{6.47}
$$

or

$$
\left.\begin{aligned}
&x_{m\pm}^{(\beta_j,j)}=x_m^{(0,j)}\left(\text{or } z_{m\pm}^{(\beta_j,j)}=z_m^{(0,j)}\right),\\
&G_{\partial\Xi_{(\alpha_j\beta_j,j)}}^{(0,j_1)}\left(\mathbf{z}_m^{(\beta_j,j)},t_{m\pm}\right)=g_1^{(\beta_j,j)}\left(\mathbf{z}_m^{(\beta_j,j)},t_{m\pm}\right)=0;\\
&(-1)^{\beta_j}G_{\partial\Xi_{(\alpha_j\beta_j,j)}}^{(1,j)}\left(\mathbf{z}_m^{(\beta_j,j)},t_{m\pm}\right)=(-1)^{\beta_j}g_2^{(\beta_j,j)}\left(\mathbf{z}_m^{(\beta_j,j)},t_{m\pm}\right)<0.
\end{aligned}\right\}\tag{6.48}
$$

Proof See the proof in Luo (2009, 2012). ■

The edge system interaction theory with higher-order singularity can be found from Luo (2009, 2012), which will not be presented herein.

References

Luo, A. C. J. (2009). *Discontinuous dynamical systems on time-varying domains*. Higher Education Press/Springer.
Luo, A. C. J. (2012). *Discontinuous dynamical systems*. Higher Education Press/Springer.

Chapter 7
Two-Identical System Synchronization

Contents

7.1 Introduction to Gyroscope System Synchronization 93
7.2 Discontinuous Dynamics 95
7.3 Synchronization and Attractivity 104
7.4 Parameter Characteristics 107
7.5 Numerical Simulations 110
References 112

In this chapter, the partial and full chaotic synchronizations of two nonlinear gyroscope systems with/without noise are investigated. From analytical conditions for synchronization and non-synchronization of two gyroscope systems, the parameter characteristic study is completed for a better understanding of the synchronization dynamics of two gyroscope dynamical systems. The boundaries of the parameter map for synchronization are determined by the onset and vanishing conditions of synchronization. The simple feedback control can make the noised gyroscope system synchronizing with chaotic behaviors of the expected gyroscope system. The methodology presented in this paper is different from other techniques for synchronization. The partial synchronization is an important phenomenon to be observed in engineering applications.

7.1 Introduction to Gyroscope System Synchronization

As in Min and Luo (2012), consider a nonlinear gyroscope system for the nonlinear gyroscope system to be

F. Min, *Discontinuous Dynamics and System Synchronization*, Nonlinear Systems and Complexity 39, https://doi.org/10.1007/978-3-031-66648-3_7

$$\ddot{\theta} + c_1\dot{\theta} + c_2\dot{\theta}^3 + h(\theta) - \beta \sin \theta = f \sin \omega t \sin \theta \tag{7.1}$$

where $\dot{\theta} = d\theta/dt$. The variable θ is rotation angle, and $f \sin \omega t \sin \theta$ is a parametric excitation. $c_1\dot{\theta}$ and $c_2\dot{\theta}^3$ are linear and nonlinear damping terms. The function $h(\theta)$ is

$$h(\theta) = \alpha^2 \frac{(1 - \cos \theta)^2}{\sin^3 \theta} \tag{7.2}$$

Setting $x_1 = \theta$ and $x_2 = \dot{\theta}$, introduce two vectors as

$$\mathbf{x} = (x_1, x_2)^{\mathrm{T}}, \text{and } \boldsymbol{\mathcal{F}}(\mathbf{x}, t) = (x_2, \mathcal{F}(\mathbf{x}, t))^{\mathrm{T}} \tag{7.3}$$

where

$$\mathcal{F}(\mathbf{x}, t) = -h(x_1) - c_1 x_2 - c_2 x_2^3 + (\beta + f_1 \sin \omega t) \sin x_1. \tag{7.4}$$

The equation of motion in Eq. 7.1 becomes

$$\dot{\mathbf{x}} = \boldsymbol{\mathcal{F}}(\mathbf{x}, t), \tag{7.5}$$

In practical application, the practical gyroscope system was noised, which cannot be exactly as the expected case in design. One expects the noised system can do exactly as what the originally expected gyroscope system can do. Consider the noised gyroscope system in Min and Luo (2012) as

$$\ddot{\theta} + c_1\dot{\theta} + c_2\dot{\theta}^3 + h(\theta) - \beta \sin \theta = f \sin \omega t \sin \theta + e(t) \tag{7.6}$$

where $e(t)$ is a noise. Without loss of generality, $e(t) = Q \cos \Omega t$ is adopted. To make the noised gyroscope system synchronize with the expected gyroscope system, a control law will be added to the noised gyroscope system in Eq. 7.6. Setting $y_1 = \theta$ and $y_2 = \dot{\theta}$, the state variables and the vector fields are introduced as

$$\mathbf{y} = (y_1, y_2)^{\mathrm{T}} \text{ and } \overline{\mathbf{F}}(\mathbf{y}, t) = (y_2, \overline{F}(\mathbf{y}, t))^{\mathrm{T}} \tag{7.7}$$

where

$$\overline{F}(\mathbf{y}, t) = -h(y_1) - c_1 y_2 - c_2 y_2^3 + (\beta + f_2 \sin \omega t) \sin y_1 + e(t). \tag{7.8}$$

Consider a control law as

$$\begin{aligned}&\mathbf{u}(\mathbf{x},\mathbf{y},t)=(u_1,u_2)^{\mathrm{T}},\\&u_1=k_1\,\mathrm{sgn}(y_1-x_1)\text{ and }u_2=k_2\,\mathrm{sgn}(y_2-x_2).\end{aligned}\tag{7.9}$$

The controlled, noised gyroscope system is

$$\dot{\mathbf{y}}=\mathbf{F}(\mathbf{y},t)=\overline{\mathbf{F}}(\mathbf{y},t)-\mathbf{u}(\mathbf{x},\mathbf{y},t).\tag{7.10}$$

7.2 Discontinuous Dynamics

As in Min and Luo (2012), under the feedback control, the noised gyroscope system becomes discontinuous. The two control laws make the controlled, noised gyroscope system in Eq. 7.10 to become four sub-dynamical systems. The corresponding vector fields for four subsystems are defined by $\mathbf{F}(\mathbf{y},t)=(F_1(\mathbf{y},t),F_2(\mathbf{y},t))^{\mathrm{T}}$, and the components in such vector fields for the controlled, noised gyroscope system are presented in the following fashion.

(i) For $y_1>x_1$ and $y_2>x_2$,

$$\begin{aligned}F_1(\mathbf{y},t)&=y_2-k_1\\F_2(\mathbf{y},t)&=-h(y_1)-c_1y_2-c_2y_2^3+(\beta+f_2\sin\omega t)\sin y_1+e(t)-k_2\end{aligned}\tag{7.11}$$

(ii) For $y_1>x_1$ and $y_2<x_2$,

$$\begin{aligned}F_1(\mathbf{y},t)&=y_2-k_1\\F_2(\mathbf{y},t)&=-h(y_1)-c_1y_2-c_2y_2^3+(\beta+f_2\sin\omega t)\sin y_1+e(t)+k_2\end{aligned}\tag{7.12}$$

(iii) For $y_1<x_1$ and $y_2<x_2$,

$$\begin{aligned}F_1(\mathbf{y},t)&=y_2+k_1\\F_2(\mathbf{y},t)&=-h(y_1)-c_1y_2-c_2y_2^3+(\beta+f_2\sin\omega t)\sin y_1+e(t)+k_2\end{aligned}\tag{7.13}$$

(iv) For $y_1<x_1$ and $y_2>x_2$,

$$\begin{aligned}F_1(\mathbf{y},t)&=y_2+k_1\\F_2(\mathbf{y},t)&=-h(y_1)-c_1y_2-c_2y_2^3+(\beta+f_2\sin\omega t)\sin y_1+e(t)-k_2\end{aligned}\tag{7.14}$$

As in Min and Luo (2012), from the above four subsystems, the corresponding domains Ω_α ($\alpha=1,2,3,4$) in phase space are defined as

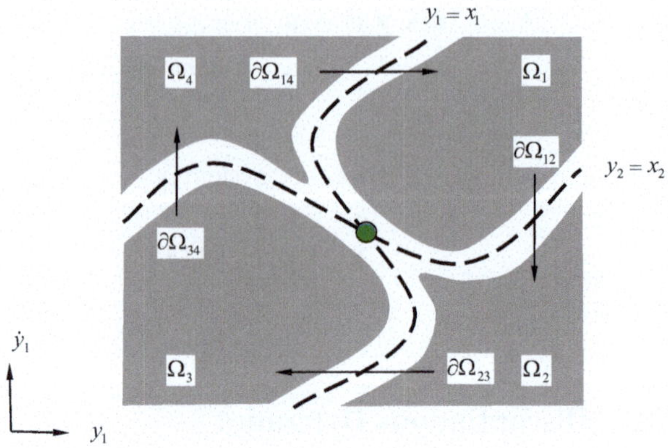

Fig. 7.1 Boundaries and domains in the absolute coordinates

$$\begin{aligned}\Omega_1 &= \{(y_1, y_2)|y_1 - x_1(t) > 0, y_2 - x_2(t) > 0\},\\ \Omega_2 &= \{(y_1, y_2)|y_1 - x_1(t) > 0, y_2 - x_2(t) < 0\},\\ \Omega_3 &= \{(y_1, y_2)|y_1 - x_1(t) < 0, y_2 - x_2(t) < 0\},\\ \Omega_4 &= \{(y_1, y_2)|y_1 - x_1(t) < 0, y_2 - x_2(t) > 0\}.\end{aligned} \tag{7.15}$$

The boundaries $\partial\Omega_{\alpha\beta}$ ($\alpha, \beta = 1, 2, 3, 4$; $\alpha \neq \beta$) of the four sub-domains are

$$\begin{aligned}\partial\Omega_{12} &= \overline{\Omega}_1 \cap \overline{\Omega}_2 = \{(y_1, y_2)|y_2 - x_2(t) = 0, y_1 - x_1(t) > 0\},\\ \partial\Omega_{23} &= \overline{\Omega}_2 \cap \overline{\Omega}_3 = \{(y_1, y_2)|y_1 - x_1(t) = 0, y_2 - x_2(t) < 0\},\\ \partial\Omega_{34} &= \overline{\Omega}_3 \cap \overline{\Omega}_4 = \{(y_1, y_2)|y_2 - x_2(t) = 0, y_1 - x_1(t) < 0\},\\ \partial\Omega_{14} &= \overline{\Omega}_1 \cap \overline{\Omega}_4 = \{(y_1, y_2)|y_1 - x_1(t) = 0, y_2 - x_2(t) > 0\}.\end{aligned} \tag{7.16}$$

Notice that $\overline{\Omega}_\alpha$ is the closure of the domain Ω_α ($\alpha = 1, 2, 3, 4$).

The sub-domains and boundaries for the controlled gyroscope system are sketched in Fig. 7.1. The corresponding domains and boundaries are labeled, and the dashed curves give the boundaries. The two boundaries are for displacement and velocity, respectively, and such boundaries are determined by the displacement and velocity of the expected gyroscope system. The intersection point of the two boundaries is labeled by a filled circular symbol.

From the defined domains, the corresponding dynamical system of the controlled, noised gyroscope system on domain Ω_α is defined as

$$\dot{\mathbf{y}}^{(\alpha)} = \mathbf{F}^{(\alpha)}\left(\mathbf{y}^{(\alpha)}, t\right) \tag{7.17}$$

where

$$\begin{aligned}
&\mathbf{F}^{(\alpha)}\left(\mathbf{y}^{(\alpha)},t\right)=\left(F_1^{(\alpha)},F_2^{(\alpha)}\right)^{\mathrm{T}},\\
&F_1^{(\alpha)}\left(\mathbf{y}^{(\alpha)},t\right)=y_2^{(\alpha)}-k_1\ \text{ for } \alpha=1,2;\\
&F_1^{(\alpha)}\left(\mathbf{y}^{(\alpha)},t\right)=y_2^{(\alpha)}+k_1\ \text{ for } \alpha=3,4;\\
&F_2^{(\alpha)}\left(\mathbf{y}^{(\alpha)},t\right)=-h\left(y_1^{(\alpha)}\right)-c_1y_2^{(\alpha)}-c_2\left(y_2^{(\alpha)}\right)^3+(\beta+f_2\sin\omega t)\sin y_1^{(\alpha)}\\
&\qquad\qquad+e(t)-k_2 \text{ for } \alpha=1,4;\\
&F_2^{(\alpha)}\left(\mathbf{y}^{(\alpha)},t\right)=-h\left(y_1^{(\alpha)}\right)-c_1y_2^{(\alpha)}-c_2\left(y_2^{(\alpha)}\right)^3+(\beta+f_2\sin\omega t)\sin y_1^{(\alpha)}\\
&\qquad\qquad+e(t)+k_2\ \text{ for } \alpha=2,3.
\end{aligned}\tag{7.18}$$

On the boundary $\partial\Omega_{\alpha\beta}=\overline{\Omega}_\alpha\cap\overline{\Omega}_\beta$, equation of motion is given by

$$\dot{\mathbf{y}}^{(\alpha\beta)}=\mathbf{F}^{(\alpha\beta)}\left(\mathbf{y}^{(\alpha\beta)},\mathbf{x}(t),t\right) \text{ and } \dot{\mathbf{x}}=\boldsymbol{\mathcal{F}}(\mathbf{x},t)\tag{7.19}$$

where

$$\begin{aligned}
&\mathbf{F}^{(\alpha\beta)}=\left(F_1^{(\alpha\beta)},F_2^{(\alpha\beta)}\right)^{\mathrm{T}},\\
&F_1^{(\alpha\beta)}\left(\mathbf{y}^{(\alpha\beta)},t\right)=y_2(t)=x_2(t) \text{ and } F_2^{(\alpha\beta)}\left(\mathbf{y}^{(\alpha\beta)},t\right)=\dot{x}_2(t).
\end{aligned}\tag{7.20}$$

with

$$\begin{aligned}
&y_1^{(\alpha\beta)}=x_1(t) \text{ and } y_2^{(\alpha\beta)}=x_2(t) \text{ on } \partial\Omega_{\alpha\beta} \text{ for } (\alpha,\beta)=(2,3),(1,4);\\
&y_1^{(\alpha\beta)}=x_1(t)+C \text{ and } y_2^{(\alpha\beta)}=x_2(t) \text{ on } \partial\Omega_{\alpha\beta} \text{ for } (\alpha,\beta)=(1,2),(3,4).
\end{aligned}\tag{7.21}$$

The boundary flow $\mathbf{x}(t)$ is given by the expected gyroscope system in Eq. 7.5. It is not easy to obtain the analytical conditions for the synchronization of two gyroscope systems. From simplicity, the relative coordinates are introduced by

$$z_1=y_1-x_1 \text{ and } \dot{z}_1\equiv z_2=y_2-x_2.\tag{7.22}$$

The domains and boundaries in the relative frame become

$$\begin{aligned}
&\Omega_1=\{(z_1,z_2)|z_1>0,z_2>0\},\\
&\Omega_2=\{(z_1,z_2)|z_1>0,z_2<0\},\\
&\Omega_3=\{(z_1,z_2)|z_1<0,z_2<0\},\\
&\Omega_4=\{(z_1,z_2)|z_1<0,z_2>0\}.
\end{aligned}\tag{7.23}$$

$$\begin{aligned}
&\partial\Omega_{12}=\{(z_1,z_2)|z_2=0,z_1>0\},\\
&\partial\Omega_{23}=\{(z_1,z_2)|z_1=0,z_2<0,\},\\
&\partial\Omega_{34}=\{(z_1,z_2)|z_2=0,z_1<0\},\\
&\partial\Omega_{14}=\{(z_1,z_2)|z_1=0,z_2>0,\}.
\end{aligned}\tag{7.24}$$

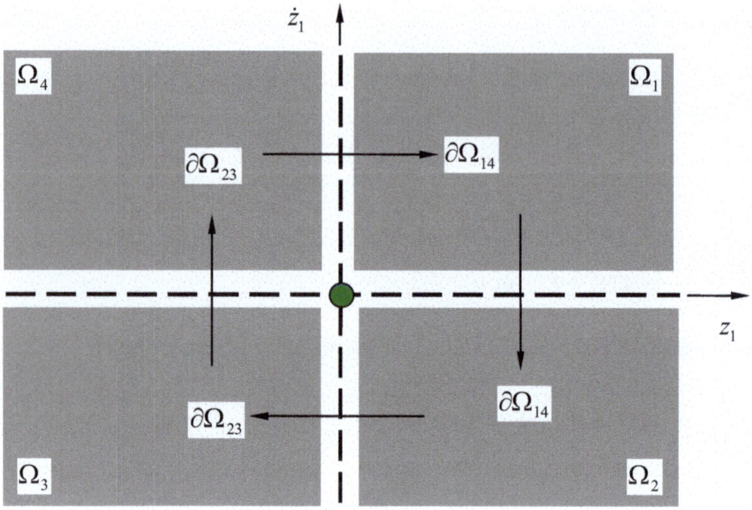

Fig. 7.2 Boundaries and domains in the relative coordinates

Similarly, the above boundaries and domains in relative coordinates are presented in Fig. 7.2, and the boundaries are straight lines, which are independent of time. The intersection point of the two boundaries is labeled by a filled circular symbol. It is very convenient to develop the analytical conditions for synchronization.

From such boundaries and domains, the analytical conditions for the aforementioned synchronization can be easily developed, and the controlled gyroscope system in the relative frame is

$$\dot{\mathbf{z}}^{(\alpha)} = \mathbf{g}^{(\alpha)}\left(\mathbf{z}^{(\alpha)}, \mathbf{x}, t\right) \text{ with } \dot{\mathbf{x}} = \mathcal{F}(\mathbf{x}, t) \tag{7.25}$$

where

$$\begin{aligned}
\mathbf{g}^{(\alpha)}\left(\mathbf{z}^{(\alpha)}, \mathbf{x}, t\right) &= \left(g_1^{(\alpha)}, g_2^{(\alpha)}\right)^{\mathrm{T}};\\
g_1^{(\alpha)}\left(\mathbf{z}^{(\alpha)}, \mathbf{x}, t\right) &= z_2^{(\alpha)} - k_1 \text{ for } \alpha = 1, 2;\\
g_1^{(\alpha)}\left(\mathbf{z}^{(\alpha)}, \mathbf{x}, t\right) &= z_2^{(\alpha)} + k_1 \text{ for } \alpha = 3, 4;\\
g_2^{(\alpha)}\left(\mathbf{z}^{(\alpha)}, \mathbf{x}, t\right) &= G\left(\mathbf{z}^{(\alpha)}, \mathbf{x}, t\right) - k_2 \text{ for } \alpha = 1, 4;\\
g_2^{(\alpha)}\left(\mathbf{z}^{(\alpha)}, \mathbf{x}, t\right) &= G\left(\mathbf{z}^{(\alpha)}, \mathbf{x}, t\right) + k_2 \text{ for } \alpha = 2, 3.
\end{aligned} \tag{7.26}$$

with

$$\begin{aligned}\mathcal{G}\left(\mathbf{z}^{(\alpha)},\mathbf{x},t\right) &= -h\left(z_1^{(\alpha)}+x_1\right)-c_1z_2^{(\alpha)}-c_2\left(\left(z_2^{(\alpha)}+x_2\right)^3-x_2^3\right)\\ &+(\beta+f_2\sin\omega t)\sin\left(z_1^{(\alpha)}+x_1\right)\\ &-(\beta+f_1\sin\omega t)\sin x_1+e(t)+h(x_1)\end{aligned} \tag{7.27}$$

The equations of motion on the boundary in the relative coordinates become

$$\dot{\mathbf{z}}^{(\alpha\beta)}=\mathbf{g}^{(\alpha\beta)}\left(\mathbf{z}^{(\alpha\beta)},\mathbf{x},t\right) \text{ with } \dot{\mathbf{x}}=\mathcal{F}(\mathbf{x},t) \tag{7.28}$$

where

$$\begin{aligned}&\mathbf{g}^{(\alpha\beta)}\left(\mathbf{z}^{(\alpha\beta)},\mathbf{x},t\right)=\left(g_1^{(\alpha\beta)},g_2^{(\alpha\beta)}\right)^{\mathrm{T}};\\ &g_1^{(\alpha\beta)}\left(\mathbf{z}^{(\alpha\beta)},\mathbf{x},t\right)=z_2=0 \text{ and } g_2^{(\alpha\beta)}\left(\mathbf{z}^{(\alpha\beta)},\mathbf{x},t\right)=0\end{aligned} \tag{7.29}$$

with

$$\begin{aligned}&z_1^{(\alpha\beta)}=0 \text{ and } z_2^{(\alpha\beta)}=0 \text{ on } \partial\Omega_{\alpha\beta} \text{ for } (\alpha,\beta)=(2,3),(1,4);\\ &z_1^{(\alpha\beta)}=C \text{ and } z_2^{(\alpha\beta)}=0 \text{ on } \partial\Omega_{\alpha\beta} \text{ for } (\alpha,\beta)=(1,2),(3,4).\end{aligned} \tag{7.30}$$

To make the noised gyroscope systems synchronize with the expected gyroscope system, from the theory of discontinuous dynamical system in Luo (2008, 2009a, 2012), the synchronization requires a sliding flow on a boundary. The non-synchronization requires a passable flow to a boundary or a flow in a domain. The desynchronization requires a source flow to a boundary. The necessary and sufficient conditions for the synchronization, non-synchronization, and desynchronization of the two gyroscope systems were presented from the synchronization theory of Luo (2009b).

To develop the analytical conditions of synchronization, G-functions are introduced via the discontinuous dynamical systems in Luo (2008, 2009a, 2012). In the relative frame, for $\mathbf{z}_m \in \partial\Omega_{ij}$ at $t=t_m$, the G-functions are

$$G_{\partial\Omega_{ij}}^{(\alpha)}(\mathbf{z}_m,\mathbf{x},t_{m\pm})=\mathbf{n}_{\partial\Omega_{ij}}^{\mathrm{T}}\cdot\left[\mathbf{g}^{(\alpha)}(\mathbf{z}_m,\mathbf{x},t_{m\pm})-\mathbf{g}^{(ij)}(\mathbf{z}_m,\mathbf{x},t_{m\pm})\right], \tag{7.31}$$

$$G_{\partial\Omega_{ij}}^{(1,\alpha)}(\mathbf{z}_m,\mathbf{x},t_{m\pm})=\mathbf{n}_{\partial\Omega_{ij}}^{\mathrm{T}}\cdot\left[D\mathbf{g}^{(\alpha)}(\mathbf{z}_m,\mathbf{x},t_{m\pm})-D\mathbf{g}^{(ij)}(\mathbf{z}_m,\mathbf{x},t_{m\pm})\right]. \tag{7.32}$$

where for $\alpha = i, j$ and $(i,j) \in \{(1,2),(2,3),(3,4),(1,4)\}$. $G^{(\alpha)}_{\partial\Omega_{ij}}(\mathbf{z}_m, \mathbf{x}, t_{m\pm})$ and $G^{(1,\alpha)}_{\partial\Omega_{ij}}(\mathbf{z}_m, \mathbf{x}, t_{m\pm})$ are the zero-order and first-order G-functions of the flow in the domain Ω_α ($\alpha \in \{i,j\}$) at the boundary $\partial\Omega_{ij}$ ($i, j = 1, 2, 3, 4$). From Eq. 7.24, the normal vectors of the boundaries in the relative frame are

$$\mathbf{n}_{\partial\Omega_{12}} = \mathbf{n}_{\partial\Omega_{34}} = (0, 1)^{\mathrm{T}} \text{ and } \mathbf{n}_{\partial\Omega_{23}} = \mathbf{n}_{\partial\Omega_{14}} = (1, 0)^{\mathrm{T}}. \tag{7.33}$$

From Eqs. 7.26–7.31, the G-functions of the flow to the domains in vicinity of the boundary becomes

$$\begin{aligned} G^{(\alpha)}_{\partial\Omega_{12}}(\mathbf{z}_m, \mathbf{x}, t_{m\pm}) &= G^{(\alpha)}_{\partial\Omega_{34}}(\mathbf{z}_m, \mathbf{x}, t_{m\pm}) = g^{(\alpha)}_2(\mathbf{z}_m, \mathbf{x}, t_{m\pm}), \\ G^{(\alpha)}_{\partial\Omega_{23}}(\mathbf{z}_m, \mathbf{x}, t_{m\pm}) &= G^{(\alpha)}_{\partial\Omega_{14}}(\mathbf{z}_m, \mathbf{x}, t_{m\pm}) = g^{(\alpha)}_1(\mathbf{z}_m, \mathbf{x}, t_{m\pm}); \end{aligned} \tag{7.34}$$

$$\begin{aligned} G^{(1,\alpha)}_{\partial\Omega_{12}}(\mathbf{z}_m, \mathbf{x}, t_{m\pm}) &= G^{(1,\alpha)}_{\partial\Omega_{34}}(\mathbf{z}_m, \mathbf{x}, t_{m\pm}) = Dg^{(\alpha)}_2(\mathbf{z}_m, \mathbf{x}, t_{m\pm}), \\ G^{(1,\alpha)}_{\partial\Omega_{23}}(\mathbf{z}_m, \mathbf{x}, t_{m\pm}) &= G^{(1,\alpha)}_{\partial\Omega_{14}}(\mathbf{z}_m, \mathbf{x}, t_{m\pm}) = Dg^{(\alpha)}_1(\mathbf{z}_m, \mathbf{x}, t_{m\pm}); \end{aligned} \tag{7.35}$$

where

$$\begin{aligned} Dg^{(\alpha)}_1\left(\mathbf{z}^{(\alpha)}, \mathbf{x}, t\right) &= g^{(\alpha)}_2\left(\mathbf{z}^{(\alpha)}, \mathbf{x}, t\right) \text{ for } \alpha = 1, 2, 3, 4; \\ Dg^{(\alpha)}_2\left(\mathbf{z}^{(\alpha)}, \mathbf{x}, t\right) &= DG\left(\mathbf{z}^{(\alpha)}, \mathbf{x}, t\right) \\ &= \left(-h_1\left(z^{(\alpha)}_1 + x_1\right) + (\beta + f_2 \sin \omega t)\cos\left(z^{(\alpha)}_1 + x_1\right)\right)F_1\left(\mathbf{z}^{(\alpha)} + \mathbf{x}, t\right) \\ &\quad - \left(c_1 + 3c_2\left(z^{(\alpha)}_2 + x_2\right)^2\right)F_2\left(\left(\mathbf{z}^{(\alpha)} + \mathbf{x}\right), t\right) \\ &\quad + \omega \cos \omega t\left(f_2 \sin\left(z^{(\alpha)}_1 + x_1\right) - f_1 \sin x_1\right) \\ &\quad - (-h_1(x_1) + (\beta + f_1 \sin \omega t)\cos x_1)F_1(\mathbf{x}, t) \\ &\quad + (c_1 + 3c_2 x_2^2)F_2(\mathbf{x}, t) - Q\Omega \sin \Omega t; \\ &\text{for } \alpha = 1, 2, 3, 4; \\ h_1\left(z^{(\alpha)}_1 + x_1\right) &= \alpha^2\left(\cos^3\left(z^{(\alpha)}_1 + x_1\right) - 4\cos^2\left(z^{(\alpha)}_1 + x_1\right) + 5\cos\left(z^{(\alpha)}_1 + x_1\right) - 2\right) / \sin^4\left(z^{(\alpha)}_1 + x_1\right), \\ h_1(x_1) &= \alpha^2(\cos^3 x_1 - 4\cos^2 x_1 + 5\cos x_1 - 2) / \sin^4 x_1, \end{aligned} \tag{7.36}$$

The G-functions of flows in domains to the boundary become

$$\begin{aligned} G^{(\alpha)}_{\partial\Omega_{12}}\left(\mathbf{z}^{(\alpha)}, \mathbf{x}, t\right) &= G^{(\alpha)}_{\partial\Omega_{34}}\left(\mathbf{z}^{(\alpha)}, \mathbf{x}, t\right) = g^{(\alpha)}_2\left(\mathbf{z}^{(\alpha)}, \mathbf{x}, t\right), \\ G^{(\alpha)}_{\partial\Omega_{23}}\left(\mathbf{z}^{(\alpha)}, \mathbf{x}, t\right) &= G^{(\alpha)}_{\partial\Omega_{14}}\left(\mathbf{z}^{(\alpha)}, \mathbf{x}, t\right) = g^{(\alpha)}_1\left(\mathbf{z}^{(\alpha)}, \mathbf{x}, t\right); \\ G^{(1,\alpha)}_{\partial\Omega_{12}}\left(\mathbf{z}^{(\alpha)}, \mathbf{x}, t\right) &= G^{(1,\alpha)}_{\partial\Omega_{34}}\left(\mathbf{z}^{(\alpha)}, \mathbf{x}, t\right) = Dg^{(\alpha)}_2\left(\mathbf{z}^{(\alpha)}, \mathbf{x}, t\right), \\ G^{(1,\alpha)}_{\partial\Omega_{23}}\left(\mathbf{z}^{(\alpha)}, \mathbf{x}, t\right) &= G^{(1,\alpha)}_{\partial\Omega_{14}}\left(\mathbf{z}^{(\alpha)}, \mathbf{x}, t\right) = Dg^{(\alpha)}_1\left(\mathbf{z}^{(\alpha)}, \mathbf{x}, t\right). \end{aligned} \tag{7.37}$$

In Eq. 7.37, for boundary $\partial\Omega_{12}$ and $\partial\Omega_{34}$, the zero-order G-function is the relative force, and the first-order G-function is the time change rate of the relative force. For boundary $\partial\Omega_{23}$ and $\partial\Omega_{14}$, the zero-order G-function is the relative velocity, and the first-order G-function is the relative force. From Luo (2008, 2009a, 2012), for the controlled, noised gyroscope system, a flow sliding on a specific boundary $\partial\Omega_{ij}$ ($(i, j) \in \{(1,2),(3,4),(2,3),(1,4)\}$) is guaranteed by

$$
\left.\begin{aligned}
G^{(1)}_{\partial\Omega_{12}}(\mathbf{z}_m,\mathbf{x},t_{m-}) = g_2^{(1)}(\mathbf{z}_m,\mathbf{x},t_{m-}) < 0,\\
G^{(2)}_{\partial\Omega_{12}}(\mathbf{z}_m,\mathbf{x},t_{m-}) = g_2^{(2)}(\mathbf{z}_m,\mathbf{x},t_{m-}) > 0
\end{aligned}\right\} \text{for } \mathbf{z}_m \in \partial\Omega_{12};
$$
$$
\left.\begin{aligned}
G^{(3)}_{\partial\Omega_{34}}(\mathbf{z}_m,\mathbf{x},t_{m-}) = g_2^{(3)}(\mathbf{z}_m,\mathbf{x},t_{m-}) > 0,\\
G^{(4)}_{\partial\Omega_{34}}(\mathbf{z}_m,\mathbf{x},t_{m-}) = g_2^{(4)}(\mathbf{z}_m,\mathbf{x},t_{m-}) < 0
\end{aligned}\right\} \text{for } \mathbf{z}_m \in \partial\Omega_{34}. \tag{7.38}
$$

$$
\left.\begin{aligned}
G^{(2)}_{\partial\Omega_{23}}(\mathbf{z}_m,\mathbf{x},t_{m-}) = g_1^{(2)}(\mathbf{z}_m,\mathbf{x},t_{m-}) < 0,\\
G^{(3)}_{\partial\Omega_{12}}(\mathbf{z}_m,\mathbf{x},t_{m-}) = g_1^{(3)}(\mathbf{z}_m,\mathbf{x},t_{m-}) > 0
\end{aligned}\right\} \text{for } \mathbf{z}_m \in \partial\Omega_{23};
$$
$$
\left.\begin{aligned}
G^{(1)}_{\partial\Omega_{14}}(\mathbf{z}_m,\mathbf{x},t_{m-}) = g_1^{(1)}(\mathbf{z}_m,\mathbf{x},t_{m-}) < 0,\\
G^{(4)}_{\partial\Omega_{14}}(\mathbf{z}_m,\mathbf{x},t_{m-}) = g_1^{(4)}(\mathbf{z}_m,\mathbf{x},t_{m-}) > 0
\end{aligned}\right\} \text{for } \mathbf{z}_m \in \partial\Omega_{14}. \tag{7.39}
$$

From Luo (2008, 2009a, 2012), for the controlled, noised gyroscope system, a flow passing through a specific boundary $\partial\Omega_{ij}$ ($(i,j) \in \{(1,2),(3,4),(2,3),(1,4)\}$) is guaranteed by

$$
\left.\begin{aligned}
G^{(1)}_{\partial\Omega_{12}}(\mathbf{z}_m,\mathbf{x},t_{m-}) = g_2^{(1)}(\mathbf{z}_m,\mathbf{x},t_{m-}) < 0,\\
G^{(2)}_{\partial\Omega_{12}}(\mathbf{z}_m,\mathbf{x},t_{m+}) = g_2^{(2)}(\mathbf{z}_m,\mathbf{x},t_{m+}) < 0
\end{aligned}\right\} \text{for } \mathbf{z}_m \in \partial\Omega_{12};
$$
$$
\left.\begin{aligned}
G^{(3)}_{\partial\Omega_{34}}(\mathbf{z}_m,\mathbf{x},t_{m-}) = g_2^{(3)}(\mathbf{z}_m,\mathbf{x},t_{m-}) > 0,\\
G^{(4)}_{\partial\Omega_{34}}(\mathbf{z}_m,\mathbf{x},t_{m+}) = g_2^{(4)}(\mathbf{z}_m,\mathbf{x},t_{m+}) > 0
\end{aligned}\right\} \text{for } \mathbf{z}_m \in \partial\Omega_{34}. \tag{7.40}
$$

$$
\left.\begin{aligned}
G^{(2)}_{\partial\Omega_{23}}(\mathbf{z}_m,\mathbf{x},t_{m-}) = g_1^{(2)}(\mathbf{z}_m,\mathbf{x},t_{m-}) < 0,\\
G^{(3)}_{\partial\Omega_{23}}(\mathbf{z}_m,\mathbf{x},t_{m+}) = g_1^{(3)}(\mathbf{z}_m,\mathbf{x},t_{m+}) < 0
\end{aligned}\right\} \text{for } \mathbf{z}_m \in \partial\Omega_{23};
$$
$$
\left.\begin{aligned}
G^{(1)}_{\partial\Omega_{14}}(\mathbf{z}_m,\mathbf{x},t_{m-}) = g_1^{(1)}(\mathbf{z}_m,\mathbf{x},t_{m-}) > 0,\\
G^{(4)}_{\partial\Omega_{14}}(\mathbf{z}_m,\mathbf{x},t_{m+}) = g_1^{(4)}(\mathbf{z}_m,\mathbf{x},t_{m+}) > 0
\end{aligned}\right\} \text{for } \mathbf{z}_m \in \partial\Omega_{14}. \tag{7.41}
$$

From Luo (2008, 2009a, 2012), for the controlled, noised gyroscope system, a flow grazing to a specific boundary $\partial\Omega_{ij}$ $((i,j)\in\{(1,2),(3,4),(2,3),(1,4)\})$ is guaranteed by

$$
\left.\begin{array}{l}
G^{(0,\alpha)}_{\partial\Omega_{12}}(\mathbf{z}_m,\mathbf{x},t_{m\pm})=g_2^{(\alpha)}(\mathbf{z}_m,\mathbf{x},t_{m\pm})=0,\\
(-1)^{\alpha}G^{(1,\alpha)}_{\partial\Omega_{12}}(\mathbf{z}_m,\mathbf{x},t_{m\pm})=(-1)^{\alpha}Dg_2^{(\alpha)}(\mathbf{z}_m,\mathbf{x},t_{m\pm})<0
\end{array}\right\}
$$
$$
\begin{array}{l}
\text{for } \mathbf{z}_m\in\partial\Omega_{12} \text{ in } \Omega_\alpha(\alpha\in\{1,2\});\\
\left.\begin{array}{l}
G^{(0,\alpha)}_{\partial\Omega_{34}}(\mathbf{z}_m,\mathbf{x},t_{m\pm})=g_2^{(\alpha)}(\mathbf{z}_m,\mathbf{x},t_{m\pm})=0,\\
(-1)^{\alpha}G^{(1,\alpha)}_{\partial\Omega_{34}}(\mathbf{z}_m,\mathbf{x},t_{m\pm})=(-1)^{\alpha}Dg_2^{(\alpha)}(\mathbf{z}_m,\mathbf{x},t_{m\pm})>0
\end{array}\right\}\\
\text{for } \mathbf{z}_m\in\partial\Omega_{34} \text{ in } \Omega_\alpha\ (\alpha\in\{3,4\});
\end{array}
\tag{7.42}
$$

$$
\begin{array}{l}
\left.\begin{array}{l}
G^{(0,\alpha)}_{\partial\Omega_{23}}(\mathbf{z}_m,\mathbf{x},t_{m\pm})=g_1^{(\alpha)}(\mathbf{z}_m,\mathbf{x},t_{m\pm})=0\\
(-1)^{\alpha}G^{(1,\alpha)}_{\partial\Omega_{23}}(\mathbf{z}_m,\mathbf{x},t_{m\pm})=(-1)^{\alpha}Dg_1^{(\alpha)}(\mathbf{z}_m,\mathbf{x},t_{m\pm})>0
\end{array}\right\}\\
\text{for } \mathbf{z}_m\in\partial\Omega_{23} \text{ in } \Omega_\alpha(\alpha\in\{2,3\});\\
\left.\begin{array}{l}
G^{(0,\alpha)}_{\partial\Omega_{14}}(\mathbf{z}_m,\mathbf{x},t_{m\pm})=g_1^{(\alpha)}(\mathbf{z}_m,\mathbf{x},t_{m\pm})=0\\
(-1)^{\alpha}G^{(1,\alpha)}_{\partial\Omega_{14}}(\mathbf{z}_m,\mathbf{x},t_{m\pm})=(-1)^{\alpha}Dg_1^{(\alpha)}(\mathbf{z}_m,\mathbf{x},t_{m\pm})<0
\end{array}\right\}\\
\text{for } \mathbf{z}_m\in\partial\Omega_{14} \text{ in } \Omega_\alpha(\alpha\in\{1,4\}).
\end{array}
\tag{7.43}
$$

From Luo (2008, 2009a, 2012), for the controlled, noised gyroscope system, the onset of a sliding flow on a specific boundary $\partial\Omega_{ij}$ $((i,j)\in\{(1,2),(3,4),(2,3),(1,4)\})$ is guaranteed by

$$
\begin{array}{l}
\left.\begin{array}{l}
G^{(0,1)}_{\partial\Omega_{12}}(\mathbf{z}_m,\mathbf{x},t_{m-})=g_2^{(1)}(\mathbf{z}_m,\mathbf{x},t_{m-})<0,\\
G^{(0,2)}_{\partial\Omega_{12}}(\mathbf{z}_m,\mathbf{x},t_{m\pm})=g_2^{(2)}(\mathbf{z}_m,\mathbf{x},t_{m\pm})=0,\\
G^{(1,2)}_{\partial\Omega_{12}}(\mathbf{z}_m,\mathbf{x},t_{m\pm})=Dg_2^{(2)}(\mathbf{z}_m,\mathbf{x},t_{m\pm})<0
\end{array}\right\}\text{ from } \Omega_1\rightarrow\partial\Omega_{12};\\
\left.\begin{array}{l}
G^{(0,3)}_{\partial\Omega_{34}}(\mathbf{z}_m,\mathbf{x},t_{m-})=g_2^{(3)}(\mathbf{z}_m,\mathbf{x},t_{m-})>0,\\
G^{(0,4)}_{\partial\Omega_{34}}(\mathbf{z}_m,\mathbf{x},t_{m\pm})=g_2^{(4)}(\mathbf{z}_m,\mathbf{x},t_{m\pm})=0,\\
G^{(1,4)}_{\partial\Omega_{34}}(\mathbf{z}_m,\mathbf{x},t_{m\pm})=Dg_2^{(4)}(\mathbf{z}_m,\mathbf{x},t_{m\pm})>0
\end{array}\right\}\text{ from } \Omega_3\rightarrow\partial\Omega_{34}.
\end{array}
\tag{7.44}
$$

$$
\left.\begin{aligned}
&G_{\partial\Omega_{23}}^{(0,2)}(\mathbf{z}_m,\mathbf{x},t_{m-})=g_1^{(2)}(\mathbf{z}_m,\mathbf{x},t_{m-})<0,\\
&G_{\partial\Omega_{23}}^{(0,3)}(\mathbf{z}_m,\mathbf{x},t_{m\pm})=g_1^{(3)}(\mathbf{z}_m,\mathbf{x},t_{m\pm})=0,\\
&G_{\partial\Omega_{23}}^{(1,3)}(\mathbf{z}_m,\mathbf{x},t_{m\pm})=Dg_1^{(3)}(\mathbf{z}_m,\mathbf{x},t_{m\pm})<0
\end{aligned}\right\}\text{ from }\Omega_2\to\partial\Omega_{23};
$$

$$
\left.\begin{aligned}
&G_{\partial\Omega_{14}}^{(0,4)}(\mathbf{z}_m,\mathbf{x},t_{m-})=g_1^{(4)}(\mathbf{z}_m,\mathbf{x},t_{m-})>0,\\
&G_{\partial\Omega_{14}}^{(0,1)}(\mathbf{z}_m,\mathbf{x},t_{m\pm})=g_1^{(1)}(\mathbf{z}_m,\mathbf{x},t_{m\pm})=0,\\
&G_{\partial\Omega_{14}}^{(1,1)}(\mathbf{z}_m,\mathbf{x},t_{m\pm})=Dg_1^{(1)}(\mathbf{z}_m,\mathbf{x},t_{m\pm})>0
\end{aligned}\right\}\text{ from }\Omega_4\to\partial\Omega_{14}.
\tag{7.45}
$$

From Luo (2008, 2009a, 2012), for the controlled, noised gyroscope system, the vanishing of a sliding flow from a specific boundary $\partial\Omega_{ij}$ ($(i,j)\in\{(1,2),(3,4),(2,3),(1,4)\}$) is guaranteed by

$$
\left.\begin{aligned}
&(-1)^{\beta}G_{\partial\Omega_{12}}^{(0,\beta)}(\mathbf{z}_m,\mathbf{x},t_{m-})=(-1)^{\beta}g_2^{(\beta)}(\mathbf{z}_m,\mathbf{x},t_{m-})>0\\
&G_{\partial\Omega_{12}}^{(0,\alpha)}(\mathbf{z}_m,\mathbf{x},t_{m\mp})=g_2^{(\alpha)}(\mathbf{z}_m,\mathbf{x},t_{m\mp})=0\\
&(-1)^{\alpha}G_{\partial\Omega_{12}}^{(1,\alpha)}(\mathbf{z}_m,\mathbf{x},t_{m\mp})=(-1)^{\alpha}Dg_2^{(\alpha)}(\mathbf{z}_m,\mathbf{x},t_{m\mp})<0,\\
&\text{for }\mathbf{z}_m\in\partial\Omega_{12};\alpha,\beta\in\{1,2\}\text{ and }\beta\neq\alpha
\end{aligned}\right\}\text{ from }\partial\Omega_{12}\to\Omega_{\alpha};
$$

$$
\left.\begin{aligned}
&(-1)^{\beta}G_{\partial\Omega_{34}}^{(0,\beta)}(\mathbf{z}_m,\mathbf{x},t_{m-})=(-1)^{\beta}g_2^{(\beta)}(\mathbf{z}_m,\mathbf{x},t_{m-})<0\\
&G_{\partial\Omega_{34}}^{(0,\alpha)}(\mathbf{z}_m,\mathbf{x},t_{m\mp})=g_2^{(\alpha)}(\mathbf{z}_m,\mathbf{x},t_{m\mp})=0\\
&(-1)^{\alpha}G_{\partial\Omega_{34}}^{(1,\alpha)}(\mathbf{z}_m,\mathbf{x},t_{m\mp})=(-1)^{\alpha}Dg_2^{(\alpha)}(\mathbf{z}_m,\mathbf{x},t_{m\mp})>0,\\
&\text{for }\mathbf{z}_m\in\partial\Omega_{34};\alpha,\beta\in\{3,4\}\text{ and }\beta\neq\alpha
\end{aligned}\right\}\text{ from }\partial\Omega_{34}\to\Omega_{\alpha};
\tag{7.46}
$$

$$
\left.\begin{aligned}
&(-1)^{\beta}G_{\partial\Omega_{23}}^{(0,\beta)}(\mathbf{z}_m,\mathbf{x},t_{m-})=(-1)^{\beta}g_1^{(\beta)}(\mathbf{z}_m,\mathbf{x},t_{m-})<0,\\
&G_{\partial\Omega_{23}}^{(0,\alpha)}(\mathbf{z}_m,\mathbf{x},t_{m\mp})=g_1^{(\alpha)}(\mathbf{z}_m,\mathbf{x},t_{m\mp})=0\\
&(-1)^{\alpha}G_{\partial\Omega_{23}}^{(1,\alpha)}(\mathbf{z}_m,\mathbf{x},t_{m\mp})=(-1)^{\alpha}Dg_1^{(\alpha)}(\mathbf{z}_m,\mathbf{x},t_{m\mp})>0\\
&\text{for }\mathbf{z}_m\in\partial\Omega_{23};\alpha,\beta\in\{2,3\}\text{ and }\beta\neq\alpha
\end{aligned}\right\}\text{ from }\partial\Omega_{23}\to\Omega_{\alpha};
$$

$$
\left.\begin{aligned}
&(-1)^{\beta}G_{\partial\Omega_{23}}^{(0,\beta)}(\mathbf{z}_m,\mathbf{x},t_{m-})=(-1)^{\beta}g_1^{(\beta)}(\mathbf{z}_m,\mathbf{x},t_{m-})>0,\\
&G_{\partial\Omega_{23}}^{(0,\alpha)}(\mathbf{z}_m,\mathbf{x},t_{m\mp})=g_1^{(\alpha)}(\mathbf{z}_m,\mathbf{x},t_{m\mp})=0\\
&(-1)^{\alpha}G_{\partial\Omega_{23}}^{(1,\alpha)}(\mathbf{z}_m,\mathbf{x},t_{m\mp})=(-1)^{\alpha}Dg_1^{(\alpha)}(\mathbf{z}_m,\mathbf{x},t_{m\mp})<0\\
&\text{for }\mathbf{z}_m\in\partial\Omega_{14};\alpha,\beta\in\{1,4\}\text{ and }\beta\neq\alpha
\end{aligned}\right\}\text{ from }\partial\Omega_{14}\to\Omega_{\alpha}.
\tag{7.47}
$$

7.3 Synchronization and Attractivity

As in Min and Luo (2012), from discontinuous dynamics of the controlled, noised gyroscope system on the boundary, the synchronization of the controlled, noised gyroscope system occurs at the intersection of the two boundaries ($\mathbf{z}_m = \mathbf{0}$), and the corresponding conditions are

$$\left.\begin{array}{l} G^{(1)}_{\partial\Omega_{14}}(\mathbf{z}_m, \mathbf{x}, t_{m-}) = g^{(1)}_1(\mathbf{z}_m, \mathbf{x}, t_{m-}) < 0, \\ G^{(1)}_{\partial\Omega_{12}}(\mathbf{z}_m, \mathbf{x}, t_{m-}) = g^{(1)}_2(\mathbf{z}_m, \mathbf{x}, t_{m-}) < 0 \end{array}\right\} \text{for } \mathbf{z}_m \in \partial\Omega_{12} \cap \partial\Omega_{14} \text{ on } \Omega_1;$$
$$\left.\begin{array}{l} G^{(2)}_{\partial\Omega_{12}}(\mathbf{z}_m, \mathbf{x}, t_{m-}) = g^{(2)}_2(\mathbf{z}_m, \mathbf{x}, t_{m-}) > 0, \\ G^{(2)}_{\partial\Omega_{23}}(\mathbf{z}_m, \mathbf{x}, t_{m-}) = g^{(2)}_1(\mathbf{z}_m, \mathbf{x}, t_{m-}) < 0 \end{array}\right\} \text{for } \mathbf{z}_m \in \partial\Omega_{12} \cap \partial\Omega_{23} \text{ on } \Omega_2;$$
$$\left.\begin{array}{l} G^{(3)}_{\partial\Omega_{23}}(\mathbf{z}_m, \mathbf{x}, t_{m-}) = g^{(3)}_1(\mathbf{z}_m, \mathbf{x}, t_{m-}) > 0, \\ G^{(3)}_{\partial\Omega_{34}}(\mathbf{z}_m, \mathbf{x}, t_{m-}) = g^{(3)}_2(\mathbf{z}_m, \mathbf{x}, t_{m-}) > 0 \end{array}\right\} \text{for } \mathbf{z}_m \in \partial\Omega_{23} \cap \partial\Omega_{34} \text{ on } \Omega_3;$$
$$\left.\begin{array}{l} G^{(4)}_{\partial\Omega_{34}}(\mathbf{z}_m, \mathbf{x}, t_{m-}) = g^{(4)}_2(\mathbf{z}_m, \mathbf{x}, t_{m-}) < 0, \\ G^{(4)}_{\partial\Omega_{14}}(\mathbf{z}_m, \mathbf{x}, t_{m-}) = g^{(4)}_1(\mathbf{z}_m, \mathbf{x}, t_{m-}) > 0 \end{array}\right\} \text{for } \mathbf{z}_m \in \partial\Omega_{34} \cap \partial\Omega_{14} \text{ on } \Omega_4. \tag{7.48}$$

From Eq. 7.26, four basic functions are defined as

$$\begin{array}{l} g_1\left(\mathbf{z}^{(\alpha)}, \mathbf{x}, t\right) \equiv g^{(\alpha)}_1\left(\mathbf{z}^{(\alpha)}, \mathbf{x}, t\right) = z^{(\alpha)}_2 - k_1 \text{ in } \Omega_\alpha \text{ for } \alpha = 1, 2; \\ g_2\left(\mathbf{z}^{(\alpha)}, \mathbf{x}, t\right) \equiv g^{(\alpha)}_1\left(\mathbf{z}^{(\alpha)}, \mathbf{x}, t\right) = z^{(\alpha)}_2 + k_1 \text{ in } \Omega_\alpha \text{ for } \alpha = 3, 4; \\ g_3\left(\mathbf{z}^{(\alpha)}, \mathbf{x}, t\right) \equiv g^{(\alpha)}_2\left(\mathbf{z}^{(\alpha)}, \mathbf{x}, t\right) = \mathcal{G}\left(\mathbf{z}^{(\alpha)}, \mathbf{x}, t\right) - k_2 \text{ in } \Omega_\alpha \text{ for } \alpha = 1, 4; \\ g_4\left(\mathbf{z}^{(\alpha)}, \mathbf{x}, t\right) \equiv g^{(\alpha)}_2\left(\mathbf{z}^{(\alpha)}, \mathbf{x}, t\right) = \mathcal{G}\left(\mathbf{z}^{(\alpha)}, \mathbf{x}, t\right) + k_2 \text{ in } \Omega_\alpha \text{ for } \alpha = 2, 3. \end{array} \tag{7.49}$$

where

$$\begin{aligned} \mathcal{G}\left(\mathbf{z}^{(\alpha)}, \mathbf{x}, t\right) = & -h\left(z^{(\alpha)}_1 + x_1\right) - c_1 z^{(\alpha)}_2 - c_2\left[\left(z^{(\alpha)}_2 + x_2\right)^3 - x_2^3\right] \\ & + (\beta + f_2 \sin\omega t)\sin\left(z^{(\alpha)}_1 + x_1\right) - (\beta + f_1 \sin\omega t)\sin x_1 + e(t) + h(x_1) \end{aligned} \tag{7.50}$$

The synchronization conditions in Eq. 7.47 become

$$\begin{aligned} g_1(\mathbf{z}_m, \mathbf{x}, t_{m-}) &= z_{2m} - k_1 < 0, \\ g_2(\mathbf{z}_m, \mathbf{x}, t_{m-}) &= z_{2m} + k_1 > 0, \\ g_3(\mathbf{z}_m, \mathbf{x}, t_{m-}) &= \mathcal{G}(\mathbf{z}_m, \mathbf{x}, t_{m-}) - k_2 < 0, \\ g_4(\mathbf{z}_m, \mathbf{x}, t_{m-}) &= \mathcal{G}(\mathbf{z}_m, \mathbf{x}, t_{m-}) + k_2 > 0. \end{aligned} \tag{7.51}$$

Letting $\mathbf{z}_m = \mathbf{0}$, the synchronization conditions of the two coupled gyroscope systems are

$$\begin{aligned} g_1(\mathbf{z}_m, \mathbf{x}, t_{m-}) &= -k_1 < 0, \\ g_2(\mathbf{z}_m, \mathbf{x}, t_{m-}) &= +k_1 > 0, \\ g_3(\mathbf{z}_m, \mathbf{x}, t_{m-}) &= e(t_{m-}) - k_2 < 0, \\ g_4(\mathbf{z}_m, \mathbf{x}, t_{m-}) &= e(t_{m-}) + k_2 > 0. \end{aligned} \tag{7.52}$$

If $k_1 > 0$ and $k_2 > 0$, the first two equations of Eq. 7.52 can be automatically satisfied, and the third and fourth equations give the synchronization invariant set as

$$-k_2 < e(t_{m-}) < k_2. \tag{7.53}$$

The conditions for vanishing of synchronization are for $\mathbf{z}^{(\alpha)}(t_{m\mp}) = \mathbf{z}_m^{(\alpha)} = \mathbf{z}_m$

$$\left.\begin{aligned} g_1(\mathbf{z}_m^{(\alpha)}, \mathbf{x}, t_{m\mp}) &= z_{2m}^{(\alpha)} - k_1 = 0, \\ Dg_1(\mathbf{z}_m^{(\alpha)}, \mathbf{x}, t_{m\mp}) &= \mathcal{G}(\mathbf{z}_m^{(\alpha)}, \mathbf{x}, t_{m\mp}) > 0, \\ g_2(\mathbf{z}_m^{(\beta)}, \mathbf{x}, t_{m-}) &= z_{2m}^{(\beta)} + k_1 > 0 \end{aligned}\right\} \text{for } (\alpha, \beta) = \{(1,4), (2,3)\} \tag{7.54}$$

from $z_{m+\varepsilon} = y_1 - x_1 > 0$, and

$$\left.\begin{aligned} g_1(\mathbf{z}_m^{(\alpha)}, \mathbf{x}, t_{m-}) &= z_{2m}^{(\alpha)} - k_1 < 0; \\ g_2(\mathbf{z}_m^{(\beta)}, \mathbf{x}, t_{m\mp}) &= z_{2m}^{(\beta)} + k_1 = 0, \\ Dg_2(\mathbf{z}_m^{(\beta)}, \mathbf{x}, t_{m\mp}) &= \mathcal{G}(\mathbf{z}_m^{(\beta)}, \mathbf{x}, t_{m\mp}) < 0 \end{aligned}\right\} \text{for } (\alpha, \beta) = \{(1,4), (2,3)\} \tag{7.55}$$

from $z_{m+\varepsilon} = y_1 - x_1 < 0$.

The conditions for vanishing of synchronization are for $\mathbf{z}^{(\alpha)}(t_{m\mp}) = \mathbf{z}_m^{(\alpha)} = \mathbf{z}_m$

$$\left.\begin{aligned} g_3(\mathbf{z}_m^{(\alpha)}, \mathbf{x}, t_{m\mp}) &= \mathcal{G}(\mathbf{z}_m^{(\alpha)}, \mathbf{x}, t_{m\mp}) - k_2 = 0, \\ Dg_3(\mathbf{z}_m^{(\alpha)}, \mathbf{x}, t_{m\mp}) &= D\mathcal{G}(\mathbf{z}_m^{(\alpha)}, \mathbf{x}, t_{m\mp}) > 0; \\ g_4(\mathbf{z}_m^{(\beta)}, \mathbf{x}, t_{m-}) &= \mathcal{G}(\mathbf{z}_m^{(\beta)}, \mathbf{x}, t_{m-}) + k_2 > 0 \end{aligned}\right\} \text{for } (\alpha, \beta) = \{(1,2), (4,3)\} \tag{7.56}$$

from $\dot{z}_{m+\varepsilon} = y_2 - x_2 > 0$, and

$$\left.\begin{aligned} g_3\big(\mathbf{z}_m^{(\alpha)}, \mathbf{x}, t_{m-}\big) &= G\big(\mathbf{z}_m^{(\alpha)}, \mathbf{x}, t_{m-}\big) - k_2 < 0; \\ g_4\big(\mathbf{z}_m^{(\beta)}, \mathbf{x}, t_{m\mp}\big) &= G\big(\mathbf{z}_m^{(\beta)}, \mathbf{x}, t_{m\mp}\big) + k_2 = 0, \\ g_4\big(\mathbf{z}_m^{(\beta)}, \mathbf{x}, t_{m\mp}\big) &= DG\big(\mathbf{z}_m^{(\beta)}, \mathbf{x}, t_{m\mp}\big) < 0 \end{aligned}\right\} \text{for } (\alpha, \beta) = \{(1,2), (4,3)\} \tag{7.57}$$

from $\dot{z}_{m+\varepsilon} = y_2 - x_2 < 0$.

The conditions for onset of synchronization are for $\mathbf{z}^{(\alpha)}(t_{m\mp}) = \mathbf{z}_m^{(\alpha)} = \mathbf{z}_m$

$$\left.\begin{aligned} g_1\big(\mathbf{z}_m^{(\alpha)}, \mathbf{x}, t_{m\pm}\big) &= z_{2m}^{(\alpha)} - k_1 = 0, \\ Dg_1\big(\mathbf{z}_m^{(\alpha)}, \mathbf{x}, t_{m\pm}\big) &= G\big(\mathbf{z}_m^{(\alpha)}, \mathbf{x}, t_{m\pm}\big) > 0, \\ g_2\big(\mathbf{z}_m^{(\beta)}, \mathbf{x}, t_{m-}\big) &= z_{2m}^{(\beta)} + k_1 > 0; \end{aligned}\right\} \text{for } (\alpha, \beta) = \{(1,4), (2,3)\} \tag{7.58}$$

from $z_{m-\varepsilon} = y_1 - x_1 > 0$, and

$$\left.\begin{aligned} g_1\big(\mathbf{z}_m^{(\alpha)}, \mathbf{x}, t_{m-}\big) &= z_{2m}^{(\alpha)} - k_1 < 0; \\ g_2\big(\mathbf{z}_m^{(\beta)}, \mathbf{x}, t_{m\pm}\big) &= z_{2m}^{(\beta)} + k_1 = 0, \\ Dg_2\big(\mathbf{z}_m^{(\beta)}, \mathbf{x}, t_{m\pm}\big) &= G\big(\mathbf{z}_m^{(\beta)}, \mathbf{x}, t_{m\pm}\big) < 0 \end{aligned}\right\} \text{for } (\alpha, \beta) = \{(1,4), (2,3)\} \tag{7.59}$$

from $z_{m+\varepsilon} = y_1 - x_1 < 0$.

The conditions for onset of synchronization are for $\mathbf{z}^{(\alpha)}(t_{m\pm}) = \mathbf{z}_m^{(\alpha)} = \mathbf{z}_m$

$$\left.\begin{aligned} g_3\big(\mathbf{z}_m^{(\alpha)}, \mathbf{x}, t_{m\pm}\big) &= G\big(\mathbf{z}_m^{(\alpha)}, \mathbf{x}, t_{m\pm}\big) - k_2 = 0, \\ Dg_3\big(\mathbf{z}_m^{(\alpha)}, \mathbf{x}, t_{m\pm}\big) &= DG\big(\mathbf{z}_m^{(\alpha)}, \mathbf{x}, t_{m\pm}\big) > 0; \\ g_4\big(\mathbf{z}_m^{(\beta)}, \mathbf{x}, t_{m-}\big) &= G\big(\mathbf{z}_m^{(\beta)}, \mathbf{x}, t_{m-}\big) + k_2 > 0 \end{aligned}\right\} \text{for } (\alpha, \beta) = \{(1,2), (4,3)\} \tag{7.60}$$

from $\dot{z}_{m-\varepsilon} = y_2 - x_2 > 0$, and

$$\left.\begin{aligned} g_3\big(\mathbf{z}_m^{(\alpha)}, \mathbf{x}, t_{m-}\big) &= G\big(\mathbf{z}_m^{(\alpha)}, \mathbf{x}, t_{m-}\big) - k_2 < 0; \\ g_4\big(\mathbf{z}_m^{(\beta)}, \mathbf{x}, t_{m\pm}\big) &= G\big(\mathbf{z}_m^{(\beta)}, \mathbf{x}, t_{m\pm}\big) + k_2 = 0, \\ g_4\big(\mathbf{z}_m^{(\beta)}, \mathbf{x}, t_{m\pm}\big) &= DG\big(\mathbf{z}_m^{(\beta)}, \mathbf{x}, t_{m\pm}\big) < 0 \end{aligned}\right\} \text{for } (\alpha, \beta) = \{(1,2), (4,3)\} \tag{7.61}$$

from $\dot{z}_{m-\varepsilon} = y_2 - x_2 < 0$.

7.4 Parameter Characteristics

From Min and Luo (2012), consider a set of parameters and initial conditions as

$$\begin{aligned}&\alpha^2 = 100, \beta = 1, c_1 = 0.5, c_2 = 0.05, \omega = 2.0, f_1 = f_2 = 35.5, \Omega = 2.0;\\&x_1 = 0.5, x_2 = 1.6 \text{ and } y_1 = 1.1, y_2 = -0.5, t_0 = 0.\end{aligned} \tag{7.62}$$

To make the noised gyroscope system synchronize with the expected gyroscope system, the control laws should be exerted. Thus, the synchronization switching scenario of the controlled, noised gyroscope system varying with control parameter k_2 is presented in Fig. 7.3 and Fig. 7.4 with $k_1 = 1$ and $Q = 1.0$ plus the other parameters and initial conditions in Eq. 7.62. Acronyms "FS," "PS," and "NS" represent *full*, *partial*, and *non*-synchronizations, respectively. Acronyms "A" and "V" denote synchronization appearance and vanishing, respectively.

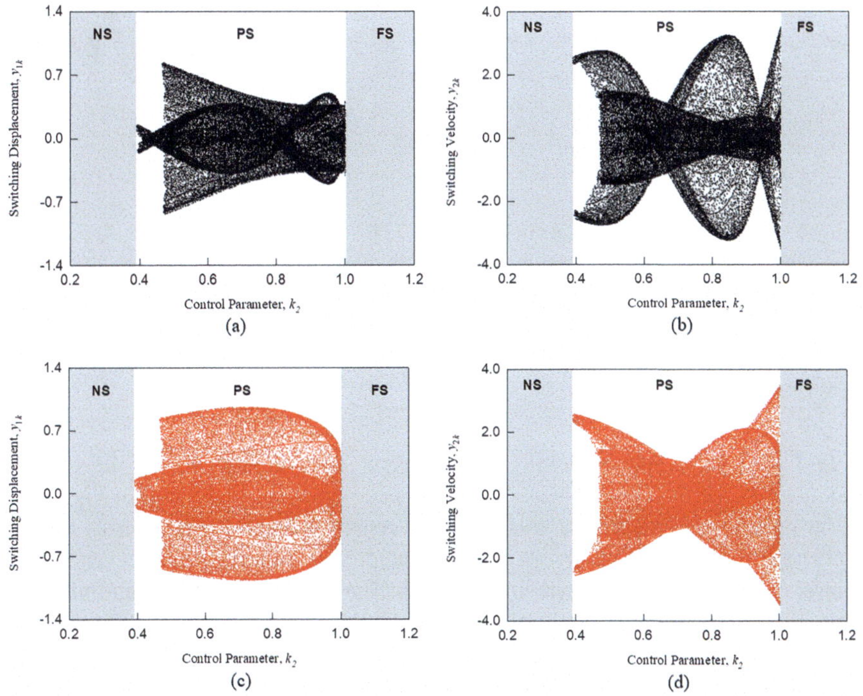

Fig. 7.3 Synchronization scenario of switching points versus control parameter k_2. Appearance: (**a**) switching displacement, (**b**) switching velocity. Disappearance: (**c**) switching displacement, (**d**) switching velocity. (Control parameter: $k_1 = 1$. System parameters: $c_1 = 0.5$, $c_2 = 0.05$, $\alpha = 10$, $\beta = 1.0$, $\omega = 2.0$, $f = 35.5$, $Q = 1.0$, $\Omega = 2.0$) (initial conditions: $x_1 = 0.5$, $x_2 = 1.6$ and $y_1 = 1.1$, $y_2 = -0.5$) (*FS* full synchronization, *PS* partial synchronization, *NS* non-synchronization)

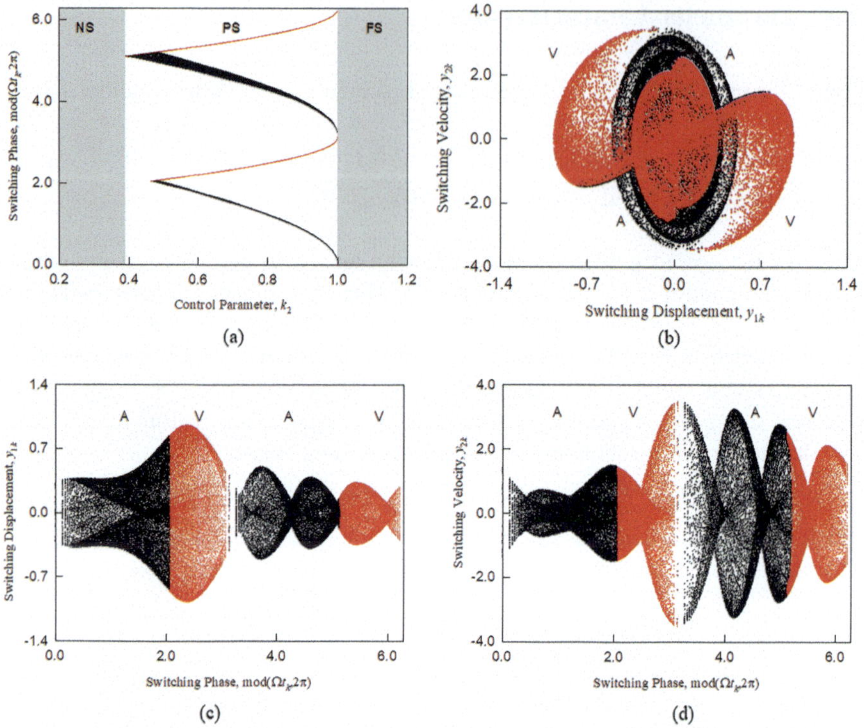

Fig. 7.4 Switching points versus control parameter k_2 for synchronization scenario. Appearance: (**a**) switching phase, (**b**) switching points in phase space, (**c**) switching displacement verse switching phase, and (**d**) switching velocity verse switching phase. (Control parameter: $k_1 = 1$. System parameters:$c_1 = 0.5$, $c_2 = 0.05$, $\alpha = 10$, $\beta = 1.0$, $\omega = 2.0, f = 35.5$. External disturbance: $Q = 1.0$, $\Omega = 2.0$). (Initial conditions:$x_1 = 0.5$, $x_2 = 1.6$ and $y_1 = 1.1$, $y_2 = -0.5$). (*FS* full synchronization, *PS* partial synchronization, *NS* non-synchronization, *A* synchronization appearance, *V* synchronization vanishing)

In Fig. 7.3a, b, the switching displacement and velocity versus the control parameter k_2 for the synchronization appearance points are presented. In Fig. 7.3c, d, the switching displacement and velocity versus the control parameter k_2 for the synchronization vanishing points are presented. The switching phase versus the control parameter k_2 is shown in Fig. 7.4a. The distribution of synchronization appearance and vanishing on the trajectory in phase plane is presented in Fig. 7.4b. The switching displacement and velocity versus the switching phase are illustrated in Fig. 7.4c, d. The switching point scenarios are more chaotic because both the expected gyroscope system and the controlled, noised gyroscope system have chaotic motions with parameters in Eq. 7.62. For the synchronization scenario, the partial synchronization of the controlled gyroscope systems with disturbance lies in $k_2 \in (0.3920, 1.000)$. If $k_2 \in (0, 0.392)$, no any synchronization can be observed. If $k_2 \in (1.000, \infty)$, the controlled, noised gyroscope system with disturbance will fully synchronize with the expected gyroscope system.

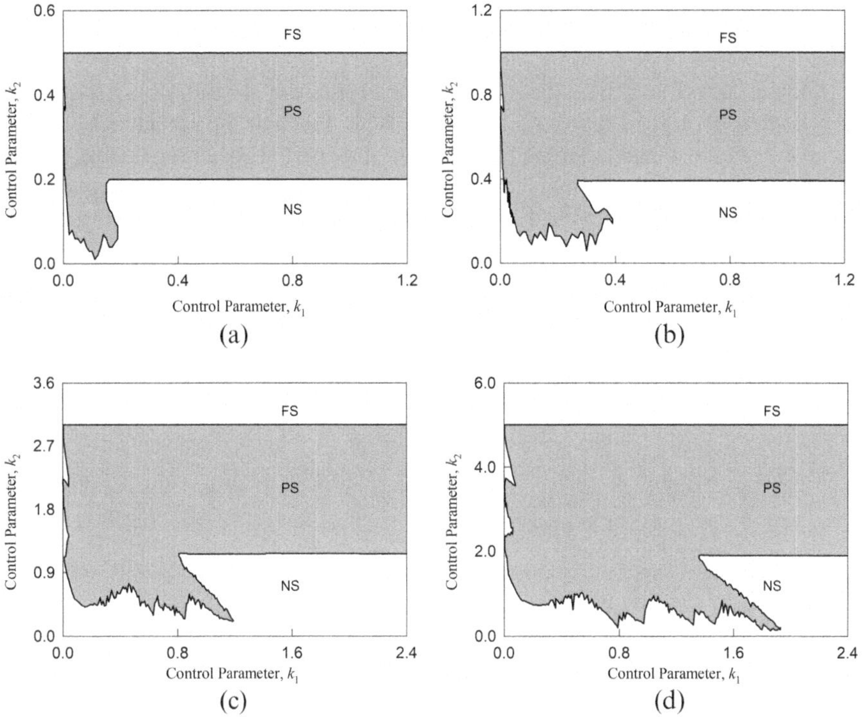

Fig. 7.5 Control parameter map of (k_1, k_2) for the synchronicity of the two coupled gyroscope dynamics with different initial conditions: (**a**) $Q = 0.5$, (**b**) $Q = 1.0$,(**a**) $Q = 3.0$,(**b**) $Q = 5.0$. (System parameters: $c_1 = 0.5$, $c_2 = 0.05$, $\alpha = 10$, $\beta = 1.0$, $\omega = 2.0$, $f = 35.5$. External noise: $\Omega = 2.0$). (Initial condition: $x_1 = 0.5$, $x_2 = 1.6$ and $y_1 = 1.1$, $y_2 = -0.5$). (*FS* full synchronization, *PS* partial synchronization, *NS* non-synchronization)

With varying control parameters of k_1 and k_2, the parameter map (k_1, k_2) for the synchronization of two coupled gyroscope systems is presented in Fig. 7.5a–d with different noise strengths of $Q = 0.5$, 1.0, 3.0, 5.0, respectively. The shade areas are partial synchronizations. Four parameter maps for the two gyroscope systems are very similar. However, the areas of partial synchronization increase with increasing noise strength. The rough boundary between the no-synchronization and partial synchronization is caused by chaotic behaviors of the expected gyroscope system. With increasing noise strength, the larger noise strength is needed for full synchronization of the controlled, noised gyroscope system with chaotic motions in the expected gyroscope system. Such a control parameter is very useful in practice. For further information, the parameter map (Q, k_2) is presented to show the relation between control parameter k_2 and noise strength. The shade areas are partial synchronizations. With increasing Q, the control parameter k_2 increases. As k_2 is less than the lower line of k_2, the noised gyroscope cannot synchronize with the expected gyroscope. This is very important in engineering application.

7.5 Numerical Simulations

In Min and Luo (Luo 2012), the system parameters are $\alpha^2 = 100$, $\beta = 1$, $c_1 = 0.5$, $c_2 = 0.05$, $\omega = 2$, $f_1 = f_2 = 35.5$, $\Omega = 2.0$. With the control parameters $k_1 = 2.4$, $k_2 = 4.0$ and excitation strength $Q = 5.0$, the partial synchronization of the controlled gyroscope system is observed in Fig. 7.6a–d with initial conditions ($x_1 = 0.6$, $x_2 = 1.8$ and $y_1 = 1.1$, $y_2 = -0.8$, $t_0 = 0$.). Two large gray circles represent two different initial conditions for master system and slave system. Acronyms "S" and "N" represent *synchronization* and *non-synchronization*. "IC" denotes *initial condition*. Hollow and filled green circular symbols are synchronization appearance and vanishing. The solid and dashed curves represent the responses of the expected and noised gyroscope systems, respectively.

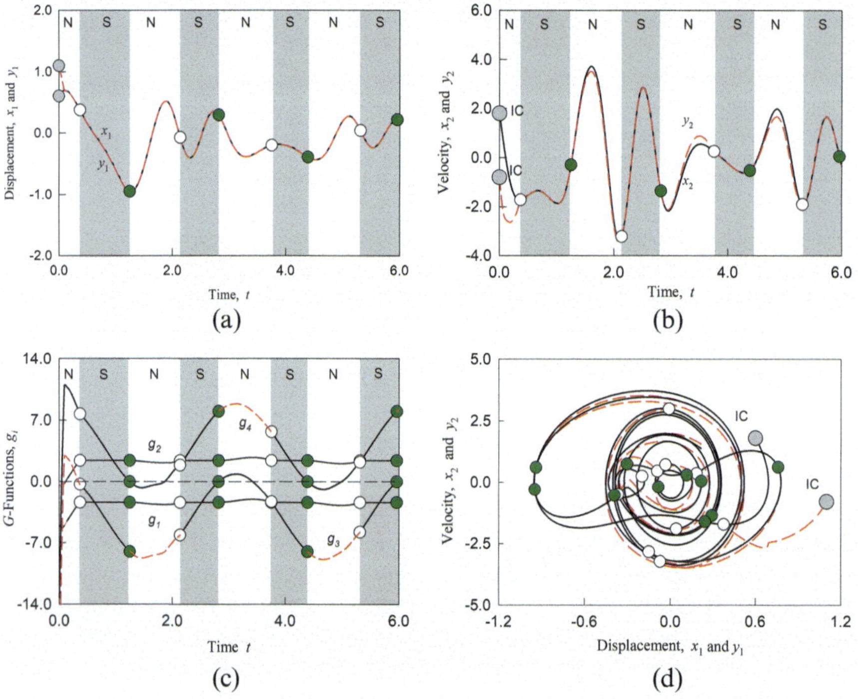

Fig. 7.6 The partial chaotic synchronization of the two gyroscope systems with different initial conditions: (**a**) displacement response, (**b**) velocity responses, (**c**) G-functions, (**d**) phase plane. (Control parameters: $k_1 = 2.4$ and $k_2 = 4.0$.) (System parameters: $c_1 = 0.5$, $c_2 = 0.05$, $\alpha = 10$, $\beta = 1.0$, $\omega = 2.0$, $f = 35.5$. External noise: $Q = 5.0$, $\Omega = 2.0$). (Initial conditions: $x_1 = 0.6$, $x_2 = 1.8$ and $y_1 = 1.1$, $y_2 = -0.8$). (*S* synchronization, *N* non-synchronization, *IC* initial conditions). Hollow and filled circular symbols are synchronization appearance (A) and vanishing (V), respectively

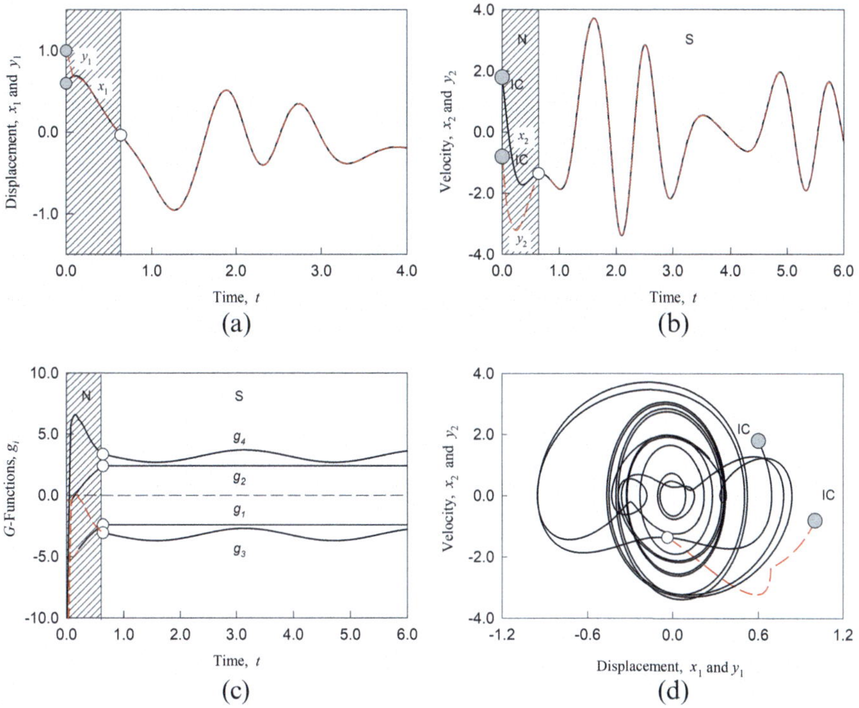

Fig. 7.7 The full chaotic synchronization of the two gyroscope dynamics with different initial conditions: (**a**) time history of velocity, (**b**) time histories of G-functions, (**c**) trajectories in phase plane for full synchronization, and (**d**) Poincare mapping section in the invariant domain. (Control parameters: $k_1 = 2.4$ and $k_2 = 3.2$). (System parameters: $c_1 = 0.5$, $c_2 = 0.05$, $\alpha = 10$, $\beta = 1.0$, $\omega = 2.0$, $f = 35.5$. External noise: $Q = 0.5$, $\Omega = 2.0$). (Initial condition: $x_1 = 0.6$, $x_2 = 1.8$ and $y_1 = 1.0$, $y_2 = -0.8$). (*S* synchronization, *N* non-synchronization, *IC*: initial conditions). Hollow circular symbols are synchronization appearance

In Fig. 7.6a, the time histories of displacements for the expected gyroscope system and the controlled, noised gyroscope system are plotted. In Fig. 7.6b, the time histories of velocity for noised and expected gyroscope systems are presented. To observe the synchronization characteristics, the time histories of G-function responses are presented in Fig. 7.6c. It is clearly observed where the noised gyroscope system will synchronize with the expected gyroscope system. The trajectories in phase planes for two coupled gyroscope systems are presented in Fig. 7.6d.

For a full synchronization of the controlled, noised gyroscope system, the control parameters $k_1 = 2.4$, $k_2 = 3.2$ and the noise strength $Q = 0.5$ are used. With initial conditions ($x_1 = 0.6$, $x_2 = 1.8$ and $y_1 = 1.0$, $y_2 = -0.8$ at $t_0 = 0.0$), the responses of the expected and noised gyroscope systems are presented in Fig. 7.7. The time histories of displacements and velocities for the expected and noised gyroscope systems are presented in Fig. 7.7a, b, respectively. For time $t \in (0, 0.6393)$, the controlled, noised gyroscope system cannot synchronize for the expected gyroscope system. For time $t \in (0.6393, \infty)$, the trajectories of two coupled gyroscope systems are identical, which implies that the expected and noised gyroscope systems are synchronized. To check the synchronization of two dynamical systems, the time histories of G-functions are presented in Fig. 7.7c. Indeed, after $t \geq 0.6393$, the two gyroscope systems are synchronized fully. The trajectories in phase plane for such a full synchronization of the expected gyroscope system and controlled, noised gyroscope system are presented in Fig. 7.7d.

References

Luo, A. C. J. (2008). A theory for flow switchability in discontinuous dynamical systems. *Nonlinear analysis: Hybrid systems, 2*(4), 1030–1061.

Luo, A. C. J. (2009a). *Discontinuous dynamical systems on time-varying domains*. HEP-Springer.

Luo, A. C. J. (2009b). A theory for synchronization of dynamical systems. *Communications in Nonlinear Science and Numerical Simulation, 14*, 1901–1951.

Luo, A. C. J. (2012). *Discontinuous dynamical systems*. Higher Education Press/Springer.

Min, F. H., & Luo, A. C. J. (2012). On parameter characteristics of chaotic synchronization in two nonlinear gyroscope systems. *Nonlinear Dynamics, 69*, 1203–1227.

Chapter 8
Chua System Synchronization and Implementation

Contents

8.1 Slave and Master Systems 113
8.2 Absolute Description 114
8.3 Relative Description 120
8.4 Analytical Synchronization Dynamics 126
8.5 Illustrations of Synchronization Dynamics 134
8.5.1 Synchronization Scenarios 135
8.5.2 Parameter Maps 136
8.5.3 Synchronization Illustrations 137
8.6 Circuit Implementation 142
References 149

In this chapter, the projective synchronization of two Chua circuits with different scrolls is presented. The analytical conditions for the projective synchronization of Chua circuits with chaotic motions are developed from the theory of discontinuous dynamical systems in Luo (2012, 2013). From those conditions, the parameter characteristics of the projective synchronization of two Chua circuit systems are discussed, and the corresponding parameter maps and the invariant domain for such projective synchronization of Chua circuits are presented. Illustrations for partial and full projective synchronizations of the Chua circuits are given. The projective synchronization of Chua circuits is implemented experimentally, which is compared with numerical results.

8.1 Slave and Master Systems

As in Min and Luo (2015), consider a modified Chua system as a master system, i.e.,

F. Min, *Discontinuous Dynamics and System Synchronization*, Nonlinear Systems and Complexity 39, https://doi.org/10.1007/978-3-031-66648-3_8

$$\begin{aligned}\dot{x}_1 &= \alpha_1[x_2 - h(x_1)],\\ \dot{x}_2 &= x_1 - x_2 + x_3,\\ \dot{x}_3 &= -\beta_1 x_2,\end{aligned} \tag{8.1}$$

where $\dot{x}_i = dx_i/dt$ (i = 1, 2, 3) and the nonlinear function $h(x_1)$ is

$$h(x_1) = ax_1 + bx_1^3. \tag{8.2}$$

The nonlinear function $h(x_1)$ is similar to the piecewise nonlinearity of the Chua circuit.

Consider a controlled Chua system with different parameters as a slave system, i.e.,

$$\begin{aligned}\dot{y}_1 &= \alpha_2[y_2 - h(y_1)] + u_1,\\ \dot{y}_2 &= y_1 - y_2 + y_3 + u_2,\\ \dot{y}_3 &= -\beta_2 y_2 + u_3,\end{aligned} \tag{8.3}$$

where $\dot{y}_i = dy_i/dt$ (i = 1, 2, 3) and the nonlinear function is also expressed as

$$h(y_1) = ay_1 + by_1^3. \tag{8.4}$$

The control laws of the projective synchronization for two Chua systems are

$$u_i = k_i \operatorname{sgn}(y_i - p_i x_i), \quad i = 1, 2, 3 \tag{8.5}$$

where k_i (i = 1, 2, 3) are controller parameters and p_i (i = 1, 2, 3) are scaling factors. The projective synchronization of the slave system with the master system can be obtained through the controlled constraints.

8.2 Absolute Description

As in Min and Luo (2015), the state variables for the master and slave systems are defined as

$$\mathbf{x} = (x_1, x_2, x_3)^T \text{ and } \mathbf{y} = (y_1, y_2, y_3)^T. \tag{8.6}$$

The master and controlled slave systems are

$$\begin{aligned}\dot{\mathbf{x}} &= \mathcal{F}(\mathbf{x}, t)\\ \dot{\mathbf{y}} &= \widetilde{\mathbf{F}}(\mathbf{y}, t) - \mathbf{u}(\mathbf{x}, \mathbf{y}, t) = \mathbf{F}(\mathbf{y}, t)\end{aligned} \tag{8.7}$$

where $\dot{\mathbf{x}} = d\mathbf{x}/dt$ and $\dot{\mathbf{y}} = d\mathbf{y}/dt$ with

$$
\begin{aligned}
&\boldsymbol{\mathcal{F}} = (\mathcal{F}_1, \mathcal{F}_2, \mathcal{F}_3)^{\mathrm{T}} \text{ and } \mathbf{F} = (F_1, F_2, F_3)^{\mathrm{T}} \\
&\mathcal{F}_1(\mathbf{x}, t) = \alpha_1[x_2 - h(x_1)], \\
&\mathcal{F}_2(\mathbf{x}, t) = x_1 - x_2 + x_3, \\
&\mathcal{F}_3(\mathbf{x}, t) = -\beta_1 x_2; \\
&F_1(\mathbf{y}, t) = \alpha_2(y_2 - h(y_1)) - u_1, \\
&F_2(\mathbf{y}, t) = y_1 - y_2 + y_3 - u_2, \\
&F_3(\mathbf{y}, t) = -\beta_2 y_2 - u_3.
\end{aligned}
\tag{8.8}
$$

The controlled Chua system with the control law in Eq. 8.5 is discontinuous. There are 8 domains and 12 boundaries with different vector fields. The eight domains Ω_α ($\alpha = 1, 2, \cdots, 8$) for the controlled Chua system are defined as

$$
\begin{aligned}
\Omega_1 &= \{(y_1, y_2, y_3) | y_1 - p_1x_1 > 0, y_2 - p_2x_2 > 0, y_3 - p_3x_3 > 0\}, \\
\Omega_2 &= \{(y_1, y_2, y_3) | y_1 - p_1x_1 > 0, y_2 - p_2x_2 > 0, y_3 - p_3x_3 < 0\}, \\
\Omega_3 &= \{(y_1, y_2, y_3) | y_1 - p_1x_1 > 0, y_2 - p_2x_2 < 0, y_3 - p_3x_3 < 0\}, \\
\Omega_4 &= \{(y_1, y_2, y_3) | y_1 - p_1x_1 > 0, y_2 - p_2x_2 < 0, y_3 - p_3x_3 > 0\}, \\
\Omega_5 &= \{(y_1, y_2, y_3) | y_1 - p_1x_1 < 0, y_2 - p_2x_2 > 0, y_3 - p_3x_3 > 0\}, \\
\Omega_6 &= \{(y_1, y_2, y_3) | y_1 - p_1x_1 < 0, y_2 - p_2x_2 > 0, y_3 - p_3x_3 < 0\}. \\
\Omega_7 &= \{(y_1, y_2, y_3) | y_1 - p_1x_1 < 0, y_2 - p_2x_2 < 0, y_3 - p_3x_3 < 0\}, \\
\Omega_8 &= \{(y_1, y_2, y_3) | y_1 - p_1x_1 < 0, y_2 - p_2x_2 > 0, y_3 - p_3x_3 > 0\}.
\end{aligned}
\tag{8.9}
$$

The 12 boundaries $\partial\Omega_{\alpha\beta}$ ($\alpha, \beta = 1, 2, \cdots, 8; \alpha \neq \beta$) of 8 domains are

$$
\begin{aligned}
\partial\Omega_{12} &= \{(y_1, y_2, y_3) | \varphi_{12} = y_3 - p_3x_3 = 0, y_1 - p_1x_1 > 0, y_2 - p_2x_2 > 0\}, \\
\partial\Omega_{23} &= \{(y_1, y_2, y_3) | \varphi_{23} = y_2 - p_2x_2 = 0, y_1 - p_1x_1 > 0, y_3 - p_3x_3 < 0\}, \\
\partial\Omega_{34} &= \{(y_1, y_2, y_3) | \varphi_{34} = y_3 - p_3x_3 = 0, y_1 - p_1x_1 > 0, y_2 - p_2x_2 < 0\}, \\
\partial\Omega_{41} &= \{(y_1, y_2, y_3) | \varphi_{41} = y_2 - p_2x_2 = 0, y_1 - p_1x_1 > 0, y_3 - p_3x_3 > 0\}, \\
\partial\Omega_{56} &= \{(y_1, y_2, y_3) | \varphi_{56} = y_3 - p_3x_3 = 0, y_1 - p_1x_1 < 0, y_2 - p_2x_2 > 0\}, \\
\partial\Omega_{67} &= \{(y_1, y_2, y_3) | \varphi_{67} = y_2 - p_2x_2 = 0, y_1 - p_1x_1 < 0, y_3 - p_3x_3 < 0\}, \\
\partial\Omega_{78} &= \{(y_1, y_2, y_3) | \varphi_{78} = y_3 - p_3x_3 = 0, y_1 - p_1x_1 < 0, y_2 - p_2x_2 < 0\}, \\
\partial\Omega_{85} &= \{(y_1, y_2, y_3) | \varphi_{85} = y_2 - p_2x_2 = 0, y_1 - p_1x_1 < 0, y_3 - p_3x_3 > 0\}, \\
\partial\Omega_{15} &= \{(y_1, y_2, y_3) | \varphi_{15} = y_1 - p_1x_1 = 0, y_2 - p_2x_2 > 0, y_3 - p_3x_3 > 0\}, \\
\partial\Omega_{26} &= \{(y_1, y_2, y_3) | \varphi_{26} = y_1 - p_1x_1 = 0, y_2 - p_2x_2 > 0, y_3 - p_3x_3 < 0\}, \\
\partial\Omega_{37} &= \{(y_1, y_2, y_3) | \varphi_{37} = y_1 - p_1x_1 = 0, y_2 - p_2x_2 < 0, y_3 - p_3x_3 < 0\}, \\
\partial\Omega_{48} &= \{(y_1, y_2, y_3) | \varphi_{48} = y_1 - p_1x_1 = 0, y_2 - p_2x_2 < 0, y_3 - p_3x_3 > 0\};
\end{aligned}
\tag{8.10}
$$

where the boundary $\partial\Omega_{\alpha\beta}$ is between Ω_α and Ω_β. The intersection boundaries of 12 boundaries form 6 edges $\wp_{l_1 l_2 l_1 l_3} = \cap_{i=1}^{4}\Omega_{l_i}$ (l_i, $= 1, 2, \cdots, 8$; $l_i \neq l_j$) which can be formed by $\mathcal{E}_{l_1 l_2 l_1 l_3} = \partial\Omega_{l_1 l_2} \cap \partial\Omega_{l_1 l_3}$, i.e.,

$$\begin{aligned}
&\mathcal{E}_{1214} = \mathcal{E}_{1223} = \mathcal{E}_{1443} = \mathcal{E}_{2343} \equiv \wp_{1234},\\
&\mathcal{E}_{5658} = \mathcal{E}_{5667} = \mathcal{E}_{5887} = \mathcal{E}_{6787} \equiv \wp_{5678},\\
&\mathcal{E}_{1215} = \mathcal{E}_{1226} = \mathcal{E}_{1556} = \mathcal{E}_{2656} \equiv \wp_{1256},\\
&\mathcal{E}_{4347} = \mathcal{E}_{4337} = \mathcal{E}_{4887} = \mathcal{E}_{3787} \equiv \wp_{3478},\\
&\mathcal{E}_{1415} = \mathcal{E}_{1448} = \mathcal{E}_{1558} = \mathcal{E}_{3885} \equiv \wp_{1458},\\
&\mathcal{E}_{2326} = \mathcal{E}_{2337} = \mathcal{E}_{2667} = \mathcal{E}_{3776} \equiv \wp_{2367};
\end{aligned} \tag{8.11}$$

where

$$\begin{aligned}
\wp_{1234} &= \{(y_1, y_2, y_3) | y_2 - p_2 x_2 = 0, y_3 - p_3 x_3 = 0, y_1 - p_1 x_1 > 0\},\\
\wp_{5678} &= \{(y_1, y_2, y_3) | y_2 - p_2 x_2 = 0, y_3 - p_3 x_3 = 0, y_1 - p_1 x_1 < 0\},\\
\wp_{1256} &= \{(y_1, y_2, y_3) | y_1 - p_1 x_1 = 0, y_3 - p_3 x_3 = 0, y_2 - p_2 x_2 > 0\},\\
\wp_{3478} &= \{(y_1, y_2, y_3) | y_1 - p_1 x_1 = 0, y_3 - p_3 x_3 = 0, y_2 - p_2 x_2 < 0\},\\
\wp_{1458} &= \{(y_1, y_2, y_3) | y_1 - p_1 x_1 = 0, y_2 - p_2 x_2 = 0, y_3 - p_3 x_3 > 0\},\\
\wp_{2367} &= \{(y_1, y_2, y_3) | y_1 - p_1 x_1 = 0, y_2 - p_2 x_2 = 0, y_3 - p_3 x_3 < 0\}
\end{aligned} \tag{8.12}$$

and the vertex of 8 domains Ω_α (or 12 boundaries $\partial\Omega_{\alpha\beta}$ ($\alpha, \beta = 1, 2, \cdots, 8; \alpha \neq \beta$) or 6 edges) is

$$\begin{aligned}
&\mathcal{E}_{l_1 l_2 l_1 l_3 l_1 l_4} = \partial\Omega_{l_1 l_2} \cap \partial\Omega_{l_1 l_3} \cap \partial\Omega_{l_1 l_4} = \mathcal{E}_{l_1 l_2 l_1 l_3} \cap \partial\Omega_{l_1 l_4}\\
&l_i = 1, 2, \cdots, 8; i = 1, 2, \cdots, 4
\end{aligned} \tag{8.13}$$

and the order of $l_1 l_2 l_1 l_3 l_1 l_4$ can be exchangeable, especially $\partial\Omega_{l_1 l_2} = \partial\Omega_{l_2 l_1}$, and so on.

$$\begin{aligned}
\mathcal{V}_{l_1 l_2 \cdots l_8} \equiv \mathcal{E}_{121415} &= \mathcal{E}_{212327} = \mathcal{E}_{234337} = \mathcal{E}_{144348}\\
&= \mathcal{E}_{155658} = \mathcal{E}_{265667} = \mathcal{E}_{376787} = \mathcal{E}_{584887}
\end{aligned} \tag{8.14}$$

where $\mathcal{V}_{l_1 l_2 \cdots l_8} = \cap_{i=1}^{8}\Omega_{l_i} (l_i = 1, 2, \cdots, 8;\ i, j = 1, 2, \cdots, 8; l_i \neq l_j \text{ if } i \neq j)$. This vertex $\mathcal{E}_{l_1 l_2 l_1 l_3 l_1 l_4}$ is in the domain Ω_{l_1}. All the vertices can be described as

$$\mathcal{E}_0 \equiv \mathcal{V}_{12345678} = \{(y_1, y_2, y_3) | y_1 - p_1 x_1 = 0, y_2 - p_2 x_2 = 0, y_3 - p_3 x_3 = 0\}. \tag{8.15}$$

On the domain Ω_α, the controlled slave system is in a vector form of

$$\dot{\mathbf{y}}^{(\alpha)} = \mathbf{F}^{(\alpha)}\left(\mathbf{y}^{(\alpha)}, t\right) \tag{8.16}$$

where

$$\mathbf{F}^{(\alpha)}\left(\mathbf{y}^{(\alpha)}, t\right) = \left(F_1^{(\alpha)}, F_2^{(\alpha)}, F_3^{(\alpha)}\right)^{\mathrm{T}}, \tag{8.17}$$

$$\begin{aligned}
&F_1^{(\alpha)}\left(\mathbf{y}^{(\alpha)}, t\right) = \alpha_2\left[y_2^{(\alpha)} - h\left(y_1^{(\alpha)}\right)\right] - k_1, \quad \text{for } \alpha = 1, 2, 3, 4;\\
&F_1^{(\alpha)}\left(\mathbf{y}^{(\alpha)}, t\right) = \alpha_2\left[y_2^{(\alpha)} - h\left(y_1^{(\alpha)}\right)\right] + k_1, \quad \text{for } \alpha = 5, 6, 7, 8;\\
&F_2^{(\alpha)}\left(\mathbf{y}^{(\alpha)}, t\right) = y_1^{(\alpha)} - y_2^{(\alpha)} + y_3^{(\alpha)} - k_2, \text{for } \alpha = 1, 2, 5, 6;\\
&F_2^{(\alpha)}\left(\mathbf{y}^{(\alpha)}, t\right) = y_1^{(\alpha)} - y_2^{(\alpha)} + y_3^{(\alpha)} + k_2, \quad \text{for } \alpha = 3, 4, 7, 8;\\
&F_3^{(\alpha)}\left(\mathbf{y}^{(\alpha)}, t\right) = -\beta_2 y_2^{(\alpha)} - k_3, \text{for } \alpha = 1, 4, 5, 8;\\
&F_3^{(\alpha)}\left(\mathbf{y}^{(\alpha)}, t\right) = -\beta_2 y_2^{(\alpha)} + k_3, \quad \text{for } \alpha = 2, 3, 6, 7.
\end{aligned} \tag{8.18}$$

Dynamical systems on the boundaries $\partial\Omega_{\alpha\beta}$ are

$$\begin{aligned}
&\dot{\mathbf{y}}^{(\alpha\beta)} = \mathbf{F}^{(\alpha\beta)}\left(\mathbf{y}^{(\alpha\beta)}, \mathbf{x}(t), t\right);\\
&\dot{\mathbf{x}} = \mathcal{F}(\mathbf{x}, t)
\end{aligned} \tag{8.19}$$

where

$$\mathbf{F}^{(\alpha\beta)} = \left(F_1^{(\alpha\beta)}, F_2^{(\alpha\beta)}, F_3^{(\alpha\beta)}\right)^{\mathrm{T}}. \tag{8.20}$$

For a passable flow from a domain Ω_α to Ω_β on the boundary $\overrightarrow{\partial\Omega}_{\alpha\beta}$, we have

$$\begin{aligned}
&F_1^{(\alpha\beta)}\left(\mathbf{y}^{(\alpha\beta)}, t\right) \in \left[F_1^{(\alpha)}\left(\mathbf{y}^{(\alpha)}, t\right), F_1^{(\beta)}\left(\mathbf{y}^{(\beta)}, t\right)\right]\\
&\text{for } (\alpha, \beta) = \{(1,5), (2,6), (3,7), (4,8)\};\\
&F_2^{(\alpha\beta)}\left(\mathbf{y}^{(\alpha\beta)}, t\right) \in \left[F_2^{(\alpha)}\left(\mathbf{y}^{(\alpha)}, t\right), F_2^{(\beta)}\left(\mathbf{y}^{(\beta)}, t\right)\right]\\
&\text{for } (\alpha, \beta) = \{(2,3), (1,4), (6,7), (8,5)\};\\
&F_3^{(\alpha\beta)}\left(\mathbf{y}^{(\alpha\beta)}, t\right) \in \left[F_3^{(\alpha)}\left(\mathbf{y}^{(\alpha)}, t\right), F_3^{(\beta)}\left(\mathbf{y}^{(\beta)}, t\right)\right]\\
&\text{for } (\alpha, \beta) = \{(1,2), (3,4), (5,6), (7,8)\}.
\end{aligned} \tag{8.21}$$

For a sliding flow on the boundary $\partial\Omega_{\alpha\beta}$, we have

$$\begin{aligned}
&F_1^{(\alpha\beta)}\left(\mathbf{y}^{(\alpha\beta)},t\right)=y_1^{(\alpha\beta)}-p_1x_1=0\\
&\text{for }\ (\alpha,\beta)=\{(1,5),(2,6),(3,7),(4,8)\};\\
&F_2^{(\alpha\beta)}\left(\mathbf{y}^{(\alpha\beta)},t\right)=y_2^{(\alpha\beta)}-p_2x_2=0\\
&\text{for }\ (\alpha,\beta)=\{(2,3),(1,4),(6,7),(8,5)\};\\
&F_3^{(\alpha\beta)}\left(\mathbf{y}^{(\alpha\beta)},t\right)=y_3^{(\alpha\beta)}-p_3x_3=0\\
&\text{for }\ (\alpha,\beta)=\{(1,2),(3,4),(5,6),(7,8)\}.
\end{aligned} \tag{8.22}$$

The other components of vector fields on the boundary $\partial\Omega_{\alpha\beta}$ are

$$\left.\begin{aligned}
&F_2^{(\alpha\beta)}(\mathbf{y}^{(\alpha\beta)},t)=y_1^{(\alpha\beta)}-y_2^{(\alpha\beta)}+y_3^{(\alpha\beta)}-k_2,\text{ for }(\alpha,\beta)=(1,5),(2,6);\\
&F_2^{(\alpha\beta)}(\mathbf{y}^{(\alpha\beta)},t)=y_1^{(\alpha\beta)}-y_2^{(\alpha\beta)}+y_3^{(\alpha\beta)}+k_2,\ \text{ for }(\alpha,\beta)=(3,7),(4,8);\\
&F_3^{(\alpha\beta)}(\mathbf{y}^{(\alpha\beta)},t)=-\beta_2y_2^{(\alpha\beta)}-k_3,\text{ for }(\alpha,\beta)=(1,5),(4,8);\\
&F_3^{(\alpha\beta)}(\mathbf{y}^{(\alpha\beta)},t)=-\beta_2y_2^{(\alpha\beta)}+k_3,\ \text{ for }(\alpha,\beta)=(3,7),(2,6);
\end{aligned}\right\} \tag{8.23}$$

$$\left.\begin{aligned}
&F_1^{(\alpha\beta)}(\mathbf{y}^{(\alpha\beta)},t)=\alpha_2\left[y_2^{(\alpha\beta)}-h\left(y_1^{(\alpha\beta)}\right)\right]-k_1,\ \text{ for }(\alpha,\beta)=(1,4),(2,3);\\
&F_1^{(\alpha\beta)}(\mathbf{y}^{(\alpha\beta)},t)=\alpha_2\left[y_2^{(\alpha\beta)}-h\left(y_1^{(\alpha\beta)}\right)\right]+k_1,\ \text{ for }(\alpha,\beta)=(5,8),(6,7);\\
&F_3^{(\alpha\beta)}(\mathbf{y}^{(\alpha\beta)},t)=-\beta_2y_2^{(\alpha\beta)}-k_3,\text{ for }(\alpha,\beta)=(1,4),(5,8);\\
&F_3^{(\alpha\beta)}(\mathbf{y}^{(\alpha\beta)},t)=-\beta_2y_2^{(\alpha\beta)}+k_3,\ \text{ for }(\alpha,\beta)=(2,3),(6,7);
\end{aligned}\right\} \tag{8.24}$$

$$\left.\begin{aligned}
&F_1^{(\alpha\beta)}(\mathbf{y}^{(\alpha\beta)},t)=\alpha_2\left[y_2^{(\alpha\beta)}-h\left(y_1^{(\alpha\beta)}\right)\right]-k_1,\ \text{ for }(\alpha,\beta)=(1,2),(3,4);\\
&F_1^{(\alpha\beta)}(\mathbf{y}^{(\alpha\beta)},t)=\alpha_2\left[y_2^{(\alpha\beta)}-h\left(y_1^{(\alpha\beta)}\right)\right]+k_1,\ \text{ for }(\alpha,\beta)=(5,6),(7,8);\\
&F_2^{(\alpha\beta)}(\mathbf{y}^{(\alpha\beta)},t)=y_1^{(\alpha\beta)}-y_2^{(\alpha\beta)}+y_3^{(\alpha\beta)}-k_2,\text{ for }(\alpha,\beta)=(1,2),(5,6);\\
&F_2^{(\alpha\beta)}(\mathbf{y}^{(\alpha\beta)},t)=y_1^{(\alpha\beta)}-y_2^{(\alpha\beta)}+y_3^{(\alpha\beta)}+k_2,\ \text{ for }(\alpha,\beta)=(3,4),(7,8).
\end{aligned}\right\} \tag{8.25}$$

On the edge $\mathcal{E}_{l_1l_2l_1l_3}$, a dynamical system is

$$\begin{aligned}
&\dot{\mathbf{y}}^{(l_1l_2l_1l_3)}=\mathbf{F}^{(l_1l_2l_1l_3)}\left(\mathbf{y}^{(l_1l_2l_1l_3)},\mathbf{x}(t),t\right);\\
&\dot{\mathbf{x}}=\boldsymbol{\mathcal{F}}(\mathbf{x},t)
\end{aligned} \tag{8.26}$$

where

$$\mathbf{F}^{(l_1 l_2 l_1 l_3)} = \left(F_1^{(l_1 l_2 l_1 l_3)}, F_2^{(l_1 l_2 l_1 l_3)}, F_3^{(l_1 l_2 l_1 l_3)}\right)^{\mathrm{T}}. \tag{8.27}$$

The edge flow to the boundary flow or domain flow can refer to Luo (2012, 2013). Only the edge flow on the edge is presented. For a sliding flow on the edge $\mathcal{E}_{l_1 l_2 l_1 l_3}$, we have

$$\left.\begin{array}{l}
F_1^{(l_1 l_2 l_1 l_3)}\left(\mathbf{y}^{(l_1 l_2 l_1 l_3)}, t\right) = y_1^{(l_1 l_2 l_1 l_3)} - p_1 x_1 = 0 \text{ on } \partial\Omega_{l_1 l_2} \\
\text{for } (l_1, l_2) = \{(1,5), (2,6), (3,7), (4,8)\}; \\
F_2^{(l_1 l_2 l_1 l_3)}\left(\mathbf{y}^{(l_1 l_2 l_1 l_3)}, t\right) = y_2^{(l_1 l_2 l_1 l_3)} - p_2 x_2 = 0 \text{ on } \partial\Omega_{l_1 l_3} \\
\text{for } (l_1, l_3) = \{(2,3), (1,4), (6,7), (8,5)\}; \\
F_3^{(l_1 l_2 l_1 l_3)}\left(\mathbf{y}^{(l_1 l_2 l_1 l_3)}, t\right) \equiv F_3^{(l_1)}\left(\mathbf{y}^{(l_1)}, t\right) = -\beta_2 y_2^{(l_1)} - k_3, \text{for } l_1 = 1,4,5,8; \\
F_3^{(l_1 l_2 l_1 l_3)}\left(\mathbf{y}^{(l_1 l_2 l_1 l_3)}, t\right) \equiv F_3^{(l_1)}\left(\mathbf{y}^{(l_1)}, t\right) = -\beta_2 y_2^{(l_1)} + k_3, \ \text{ for } l_1 = 2,3,6,7;
\end{array}\right\} \tag{8.28}$$

$$\left.\begin{array}{l}
F_1^{(l_1 l_2 l_1 l_3)}\left(\mathbf{y}^{(l_1 l_2 l_1 l_3)}, t\right) = y_1^{(l_1 l_2 l_1 l_3)} - p_1 x_1 = 0 \text{ on } \partial\Omega_{l_1 l_2} \\
\text{for } (l_1, l_2) = \{(1,5), (2,6), (3,7), (4,8)\}; \\
F_3^{(l_1 l_2 l_1 l_3)}\left(\mathbf{y}^{(l_1 l_2 l_1 l_3)}, t\right) = y_3^{(l_1 l_2 l_1 l_3)} - p_3 x_3 = 0 \text{ on } \partial\Omega_{l_1 l_3} \\
\text{for } (l_1, l_3) = \{(1,2), (3,4), (5,6), (7,8)\}. \\
F_2^{(l_1 l_2 l_1 l_3)}\left(\mathbf{y}^{(l_1 l_2 l_1 l_3)}, t\right) \equiv F_2^{(l_1)}\left(\mathbf{y}^{(l_1)}, t\right) = y_1^{(l_1)} - y_2^{(l_1)} + y_3^{(l_1)} - k_2, \text{for } l_1 = 1,2,5,6; \\
F_2^{(l_1 l_2 l_1 l_3)}\left(\mathbf{y}^{(l_1 l_2 l_1 l_3)}, t\right) \equiv F_2^{(l_1)}\left(\mathbf{y}^{(l_1)}, t\right) = y_1^{(l_1)} - y_2^{(l_1)} + y_3^{(l_1)} + k_2, \ \text{ for } l_1 = 3,4,7,8;
\end{array}\right\} \tag{8.29}$$

$$\left.\begin{array}{l}
F_2^{(l_1 l_2 l_1 l_3)}\left(\mathbf{y}^{(l_1 l_2 l_1 l_3)}, t\right) = y_2^{(l_1 l_2 l_1 l_3)} - p_2 x_2 = 0 \text{ on } \partial\Omega_{l_1 l_2} \\
\text{for } (l_1, l_2) = \{(2,3), (1,4), (6,7), (8,5)\}; \\
F_3^{(l_1 l_2 l_1 l_3)}\left(\mathbf{y}^{(l_1 l_2 l_1 l_3)}, t\right) = y_3^{(l_1 l_2 l_1 l_3)} - p_3 x_3 = 0 \text{ on } \partial\Omega_{l_1 l_3} \\
\text{for } (l_1, l_3) = \{(1,2), (3,4), (5,6), (7,8)\}; \\
F_1^{(l_1 l_2 l_1 l_3)}\left(\mathbf{y}^{(l_1 l_2 l_1 l_3)}, t\right) \equiv F_1^{(l_1)}\left(\mathbf{y}^{(l_1)}, t\right) = \alpha_2\left[y_2^{(l_1)} - h\left(y_1^{(l_1)}\right)\right] - k_1, \ \text{ for } l_1 = 1,2,3,4; \\
F_1^{(l_1 l_2 l_1 l_3)}\left(\mathbf{y}^{(l_1 l_2 l_1 l_3)}, t\right) \equiv F_1^{(l_1)}\left(\mathbf{y}^{(l_1)}, t\right) = \alpha_2\left[y_2^{(l_1)} - h\left(y_1^{(l_1)}\right)\right] - k, \ \text{ for } l_1 = 5,6,7,8.
\end{array}\right\} \tag{8.30}$$

On the vertex $\mathcal{E}_0$, dynamical system is

$$\begin{array}{l}
\dot{\mathbf{y}}^{(0)} = \mathbf{F}^{(0)}\left(\mathbf{y}^{(0)}, \mathbf{x}(t), t\right); \\
\dot{\mathbf{x}} = \mathcal{F}(\mathbf{x}, t)
\end{array} \tag{8.31}$$

where

$$
\begin{aligned}
&\mathbf{F}^{(0)} = \left(F_1^{(0)}, F_2^{(0)}, F_3^{(0)}\right)^{\mathrm{T}}, \\
&F_1^{(0)} = y_1^{(0)} - p_1 x_1 = 0, \\
&F_2^{(0)} = y_2^{(0)} - p_2 x_2 = 0, \\
&F_2^{(0)} = y_3^{(0)} - p_3 x_3 = 0.
\end{aligned} \tag{8.32}
$$

The vertex flow to the edge flow, or boundary flow, or domain flows will not be discussed herein, which can refer to Luo (2013). Only the vertex flow on the vertex is presented.

8.3 Relative Description

As in Min and Luo (2015), for simplicity, relative coordinates are introduced as

$$z_1 = y_1 - p_1 x_1, z_2 = y_2 - p_2 x_2, \text{and } z_3 = y_3 - p_3 x_3. \tag{8.33}$$

The corresponding domains, boundaries, edges, and synchronization corner in the relative frame can be expressed. The domains in the relative frame are

$$
\begin{aligned}
&{}^r\Omega_{r1} = \{(z_1, z_2, z_3) | z_1 > 0, z_2 > 0, z_3 > 0\}, \\
&{}^r\Omega_2 = \{(z_1, z_2, z_3) | z_1 > 0, z_2 > 0, z_3 < 0\}, \\
&{}^r\Omega_3 = \{(z_1, z_2, z_3) | z_1 > 0, z_2 < 0, z_3 < 0\}, \\
&{}^r\Omega_4 = \{(z_1, z_2, z_3) | z_1 > 0, z_2 < 0, z_3 > 0\}, \\
&{}^r\Omega_5 = \{(z_1, z_2, z_3) | z_1 < 0, z_2 > 0, z_3 > 0\}, \\
&{}^r\Omega_6 = \{(z_1, z_2, z_3) | z_1 < 0, z_2 > 0, z_3 < 0\}, \\
&{}^r\Omega_7 = \{(z_1, z_2, z_3) | z_1 < 0, z_2 < 0, z_3 < 0\}, \\
&{}^r\Omega_8 = \{(z_1, z_2, z_3) | z_1 < 0, z_2 < 0, z_3 > 0\}.
\end{aligned} \tag{8.34}
$$

The boundaries in the relative frame are

$$
\begin{aligned}
&{}^r\partial\Omega_{12} = \{(z_1, z_2, z_3) | \varphi_{12} = z_3 = 0, z_1 > 0, z_2 > 0\}, \\
&{}^r\partial\Omega_{23} = \{(z_1, z_2, z_3) | \varphi_{23} = z_2 = 0, z_1 > 0, z_3 < 0\}, \\
&{}^r\partial\Omega_{34} = \{(z_1, z_2, z_3) | \varphi_{34} = z_3 = 0, z_1 > 0, z_2 < 0\}, \\
&{}^r\partial\Omega_{41} = \{(z_1, z_2, z_3) | \varphi_{41} = z_2 = 0, z_1 > 0, z_3 > 0\};
\end{aligned} \tag{8.35}
$$

$$\begin{aligned}
{}^r\partial\Omega_{56} &= \{(z_1, z_2, z_3)|\varphi_{56} = z_3 = 0, z_1 < 0, z_2 > 0\},\\
{}^r\partial\Omega_{67} &= \{(z_1, z_2, z_3)|\varphi_{67} = z_2 = 0, z_1 < 0, z_3 < 0\},\\
{}^r\partial\Omega_{78} &= \{(z_1, z_2, z_3)|\varphi_{78} = z_3 = 0, z_1 < 0, z_2 < 0\},\\
{}^r\partial\Omega_{85} &= \{(z_1, z_2, z_3)|\varphi_{85} = z_2 = 0, z_1 < 0, z_3 > 0\};
\end{aligned} \tag{8.36}$$

$$\begin{aligned}
{}^r\partial\Omega_{15} &= \{(z_1, z_2, z_3)|\varphi_{15} = z_1 = 0, z_2 > 0, z_3 > 0\},\\
{}^r\partial\Omega_{26} &= \{(z_1, z_2, z_3)|\varphi_{26} = z_1 = 0, z_2 > 0, z_3 < 0\},\\
{}^r\partial\Omega_{37} &= \{(z_1, z_2, z_3)|\varphi_{37} = z_1 = 0, z_2 < 0, z_3 < 0\},\\
{}^r\partial\Omega_{48} &= \{(z_1, z_2, z_3)|\varphi_{48} = z_1 = 0, z_2 < 0, z_3 > 0\}.
\end{aligned} \tag{8.37}$$

The edges ${}^r\wp_{l_1l_2l_1l_3} = \cap_{i=1}^4 {}^r\Omega_{l_i}$ (l_i, = 1, 2, ⋯, 8; $l_i \neq l_j$) in the relative frame can be formed by ${}^r\mathcal{E}_{l_1l_2l_1l_3} = {}^r\partial\Omega_{l_1l_2} \cap {}^r\partial\Omega_{l_1l_3}$, i.e.,

$$\begin{aligned}
{}^r\mathcal{E}_{1214} &= {}^r\mathcal{E}_{1223} = {}^r\mathcal{E}_{1443} = {}^r\mathcal{E}_{2343} \equiv {}^r\wp_{1234},\\
{}^r\mathcal{E}_{5658} &= {}^r\mathcal{E}_{5667} = {}^r\mathcal{E}_{5887} = {}^r\mathcal{E}_{6787} \equiv {}^r\wp_{5678},\\
{}^r\mathcal{E}_{1215} &= {}^r\mathcal{E}_{1226} = {}^r\mathcal{E}_{1556} = {}^r\mathcal{E}_{2656} \equiv {}^r\wp_{1256},\\
{}^r\mathcal{E}_{4347} &= {}^r\mathcal{E}_{4337} = {}^r\mathcal{E}_{4887} = {}^r\mathcal{E}_{3787} \equiv {}^r\wp_{3478},\\
{}^r\mathcal{E}_{1415} &= {}^r\mathcal{E}_{1448} = {}^r\mathcal{E}_{1558} = {}^r\mathcal{E}_{3885} \equiv {}^r\wp_{1458},\\
{}^r\mathcal{E}_{2326} &= {}^r\mathcal{E}_{2337} = {}^r\mathcal{E}_{2667} = {}^r\mathcal{E}_{3776} \equiv {}^r\wp_{2367}.
\end{aligned} \tag{8.38}$$

The edges in the relative frame becomes

$$\begin{aligned}
{}^r\wp_{1234} &= \{(z_1, z_2, z_3)|z_2 = 0, z_3 = 0, z_1 > 0\},\\
{}^r\wp_{5678} &= \{(z_1, z_2, z_3)|z_2 = 0, z_3 = 0, z_1 < 0\},\\
{}^r\wp_{1256} &= \{(z_1, z_2, z_3)|z_1 = 0, z_3 = 0, z_2 > 0\},\\
{}^r\wp_{3478} &= \{(z_1, z_2, z_3)|z_1 = 0, z_3 = 0, z_2 < 0\},\\
{}^r\wp_{1458} &= \{(z_1, z_2, z_3)|z_1 = 0, z_2 = 0, z_3 > 0\},\\
{}^r\wp_{2367} &= \{(z_1, z_2, z_3)|z_1 = 0, z_2 = 0, z_3 < 0\}.
\end{aligned} \tag{8.39}$$

The corner of 8 domains ${}^r\Omega_\alpha$ with 12 boundaries ${}^r\partial\Omega_{\alpha\beta}$ ($\alpha, \beta = 1, 2, \cdots, 8; \alpha \neq \beta$) and 6 edges in the relative frame is

$$\begin{aligned}
&{}^r\mathcal{E}_{l_1l_2l_1l_3l_1l_4} = {}^r\partial\Omega_{l_1l_2} \cap {}^r\partial\Omega_{l_1l_3} \cap {}^r\partial\Omega_{l_1l_4} = {}^r\mathcal{E}_{l_1l_2l_1l_3} \cap {}^r\partial\Omega_{l_1l_4}\\
&l_i = 1, 2, \cdots, 8; i = 1, 2, \cdots, 4
\end{aligned} \tag{8.40}$$

and

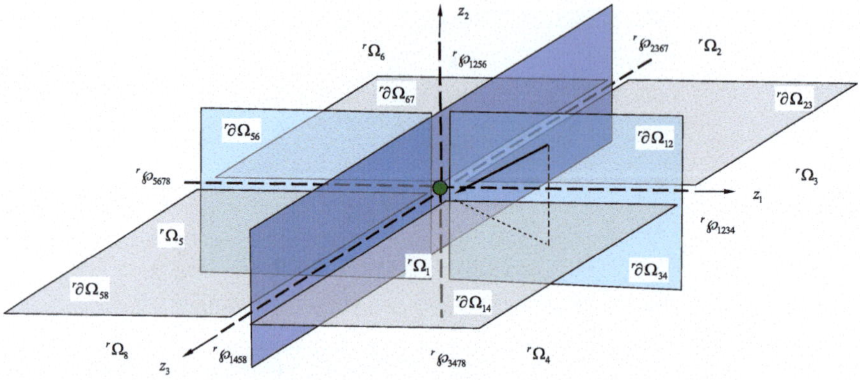

Fig. 8.1 Phase plane partitions and boundaries in relative coordinates

$$\begin{aligned} {}^{r}\mathcal{V}_{l_1 l_2 \cdots l_8} \equiv {}^{r}\mathcal{E}_{121415} &= {}^{r}\mathcal{E}_{212327} = {}^{r}\mathcal{E}_{234337} = {}^{r}\mathcal{E}_{144348} \\ &= {}^{r}\mathcal{E}_{155658} = {}^{r}\mathcal{E}_{265667} = {}^{r}\mathcal{E}_{376787} = {}^{r}\mathcal{E}_{584887} \end{aligned} \tag{8.41}$$

where ${}^{r}\mathcal{V}_{l_1 l_2 \cdots l_8} = \cap_{i=1}^{8} {}^{r}\Omega_{l_i} (l_i = 1, 2, \cdots, 8;\ i, j = 1, 2, \cdots, 8;\ l_i \neq l_j \text{ if } i \neq j)$. This vertex ${}^{r}\mathcal{E}_{l_1 l_2 l_1 l_3 l_1 l_4}$ is in the domain Ω_{l_1}. For all the vertices, it can be described as

$${}^{r}\mathcal{E}_0 = {}^{r}\mathcal{V}_{12345678} = \{(z_1, z_2, z_3) | z_1 = 0, z_2 = 0, z_3 = 0\}. \tag{8.42}$$

From the above definition, the boundaries, edges, and vertex in the relative coordinates are independent of time, as shown in Fig. 8.1. For such domains, boundaries, edges, and vertex, the analytical conditions for projective synchronization of the controlled slave system and master system can be developed from the theory for discontinuous dynamical systems. The controlled Chua circuit in the relative coordinates becomes

$$\dot{\mathbf{z}}^{(\alpha)} = \mathbf{g}^{(\alpha)}\left(\mathbf{z}^{(\alpha)}, \mathbf{x}, t\right) \text{ with } \dot{\mathbf{x}} = \mathcal{F}(\mathbf{x}, t) \tag{8.43}$$

where

$$\mathbf{g}^{(\alpha)}\left(\mathbf{z}^{(\alpha)}, \mathbf{x}, t\right) = \left(g_1^{(\alpha)}, g_2^{(\alpha)}, g_3^{(\alpha)}\right)^{\mathrm{T}}. \tag{8.44}$$

The corresponding components are

$$
\begin{aligned}
g_1^{(\alpha)}\left(\mathbf{z}^{(\alpha)},\mathbf{x},t\right) &= \alpha_2\left[z_2^{(\alpha)}+p_2x_2-h\left(z_1^{(\alpha)}+p_1x_1\right)\right]-p_1\alpha_1[x_2-h(x_1)]-k_1\\
&\text{for } \alpha = 1,2,3,4;\\
g_1^{(\alpha)}\left(\mathbf{z}^{(\alpha)},\mathbf{x},t\right) &= \alpha_2\left[z_2^{(\alpha)}+p_2x_2-h\left(z_1^{(\alpha)}+p_1x_1\right)\right]-p_1\alpha_1[x_2-h(x_1)]+k_1\\
&\text{for } \alpha = 5,6,7,8;\\
g_2^{(\alpha)}\left(\mathbf{z}^{(\alpha)},\mathbf{x},t\right) &= z_1^{(\alpha)}+p_1x_1-\left(z_2^{(\alpha)}+p_2x_2\right)+z_3^{(\alpha)}+p_3x_3-p_2(x_1-x_2+x_3)-k_2\\
&\text{for } \alpha = 1,2,5,6;\\
g_2^{(\alpha)}\left(\mathbf{z}^{(\alpha)},\mathbf{x},t\right) &= z_1^{(\alpha)}+p_1x_1-\left(z_2^{(\alpha)}+p_2x_2\right)+z_3^{(\alpha)}+p_3x_3-p_2(x_1-x_2+x_3)+k_2\\
&\text{for } \alpha = 3,4,7,8;\\
g_3^{(\alpha)}\left(\mathbf{z}^{(\alpha)},\mathbf{x},t\right) &= -\beta_2\left(z_2^{(\alpha)}+p_2x_2\right)+p_3(\beta_1x_2)-k_3, \quad \text{for } \alpha=1,4,5,8;\\
g_3^{(\alpha)}\left(\mathbf{z}^{(\alpha)},\mathbf{x},t\right) &= -\beta_2\left(z_2^{(\alpha)}+p_2x_2\right)+p_3(\beta_1x_2)+k_3, \quad \text{for } \alpha=2,3,6,7;
\end{aligned}
\tag{8.45}
$$

with

$$
h\left(z_1^{(\alpha)}+p_1x_1\right)=a\left(z_1^{(\alpha)}+p_1x_1\right)+b\left(z_1^{(\alpha)}+p_1x_1\right)^3 \tag{8.46}
$$

The equations of motion on the boundary in the relative coordinates become

$$
\dot{\mathbf{z}}^{(\alpha\beta)}=\mathbf{g}^{(\alpha\beta)}\left(\mathbf{z}^{(\alpha\beta)},\mathbf{x},t\right) \text{ with } \dot{\mathbf{x}}=\boldsymbol{\mathcal{F}}(\mathbf{x},t); \tag{8.47}
$$

where

$$
\mathbf{g}^{(\alpha\beta)}\left(\mathbf{z}^{(\alpha\beta)},\mathbf{x},t\right)=\left(g_1^{(\alpha\beta)},g_2^{(\alpha\beta)},g_3^{(\alpha\beta)}\right)^{\mathrm{T}}. \tag{8.48}
$$

For a passable flow from a domain Ω_α to Ω_β on the boundary $\partial\Omega_{\alpha\beta}$, we have

$$
\begin{aligned}
&g_1^{(\alpha\beta)} \in \left[g_1^{(\alpha)}\left(\mathbf{z}^{(\alpha)},\mathbf{x},t\right),g_1^{(\beta)}\left(\mathbf{z}^{(\beta)},\mathbf{x},t\right)\right]\\
&\text{for } (\alpha,\beta)=\{(1,5),(2,6),(3,7),(4,8)\};\\
&g_2^{(\alpha\beta)} \in \left[g_2^{(\alpha)}\left(\mathbf{z}^{(\alpha)},\mathbf{x},t\right),g_2^{(\beta)}\left(\mathbf{z}^{(\beta)},\mathbf{x},t\right)\right]\\
&\text{for } (\alpha,\beta)=\{(2,3),(1,4),(6,7),(8,5)\};\\
&g_3^{(\alpha\beta)} \in \left[g_3^{(\alpha)}\left(\mathbf{z}^{(\alpha)},\mathbf{x},t\right),g_2^{(\beta)}\left(\mathbf{z}^{(\beta)},\mathbf{x},t\right)\right]\\
&\text{for } (\alpha,\beta)=\{(1,2),(3,4),(5,6),(7,8)\}.
\end{aligned}
\tag{8.49}
$$

For a sliding flow on the boundary $\partial\Omega_{\alpha\beta}$, we have

$$
\begin{aligned}
&g_1^{(\alpha\beta)} = z_1^{(\alpha\beta)} = 0\\
&\text{for } (\alpha,\beta) = \{(1,5),(2,6),(3,7),(4,8)\};\\
&g_2^{(\alpha\beta)} = z_2^{(\alpha\beta)} = 0\\
&\text{for } (\alpha,\beta) = \{(2,3),(1,4),(6,7),(8,5)\};\\
&g_3^{(\alpha\beta)} = z_3^{(\alpha\beta)} = 0\\
&\text{for } (\alpha,\beta) = \{(1,2),(3,4),(5,6),(7,8)\}.
\end{aligned} \tag{8.50}
$$

The other components of vector fields on the boundary $\partial\Omega_{\alpha\beta}$ are

$$
\left.
\begin{aligned}
&g_2^{(\alpha\beta)} = z_1^{(\alpha\beta)} + p_1x_1 - \left(z_2^{(\alpha\beta)} + p_2x_2\right) + z_3^{(\alpha\beta)} + p_3x_3 - p_2(x_1 - x_2 + x_3) - k_2,\\
&\text{for } (\alpha,\beta) = (1,5),(2,6);\\
&g_2^{(\alpha\beta)} = z_1^{(\alpha\beta)} + p_1x_1 - \left(z_2^{(\alpha\beta)} + p_2x_2\right) + z_3^{(\alpha\beta)} + p_3x_3 - p_2(x_1 - x_2 + x_3) + k_2,\\
&\text{for } (\alpha,\beta) = (3,7),(4,8);\\
&g_3^{(\alpha\beta)} = -\beta_2\left(z_2^{(\alpha\beta)} + p_2x_2\right) + \beta_1p_3x_2 - k_3, \text{for } (\alpha,\beta) = (1,5),(4,8);\\
&g_3^{(\alpha\beta)} = -\beta_2\left(z_2^{(\alpha\beta)} + p_2x_2\right) + \beta_1p_3x_2 - k_3, \ \text{for } (\alpha,\beta) = (2,6),(3,7).
\end{aligned}
\right\} \tag{8.51}
$$

$$
\left.
\begin{aligned}
g_1^{(\alpha\beta)} &= \alpha_2\left[z_2^{(\alpha\beta)} + p_2x_2 - h\left(z_1^{(\alpha\beta)} + p_1x_1\right)\right] - p_1\alpha_1[x_2 - h(x_1)] - k_1\\
\text{for } (\alpha,\beta) &= (2,3),(1,4);\\
g_1^{(\alpha\beta)} &= \alpha_2\left[z_2^{(\alpha\beta)} + p_2x_2 - h\left(z_1^{(\alpha\beta)} + p_1x_1\right)\right] - p_1\alpha_1[x_2 - h(x_1)] + k_1\\
\text{for } (\alpha,\beta) &= (6,7),(5,8);\\
g_3^{(\alpha\beta)} &= -\beta_2\left(z_2^{(\alpha\beta)} + p_2x_2\right) + \beta_1p_3x_2 - k_3, \text{for } (\alpha,\beta) = (1,4),(5,8);\\
g_3^{(\alpha\beta)} &= -\beta_2\left(z_2^{(\alpha\beta)} + p_2x_2\right) + \beta_1p_3x_2 - k_3, \ \text{for } \alpha,\beta = (2,3),(6,7).
\end{aligned}
\right\} \tag{8.52}
$$

$$
\left.
\begin{aligned}
g_1^{(\alpha\beta)} &= \alpha_2\left[z_2^{(\alpha\beta)} + p_2x_2 - h\left(z_1^{(\alpha\beta)} + p_1x_1\right)\right] - p_1\alpha_1[x_2 - h(x_1)] - k_1,\\
\text{for } (\alpha,\beta) &= (1,2),(3,4);\\
g_1^{(\alpha\beta)} &= \alpha_2\left[z_2^{(\alpha\beta)} + p_2x_2 - h\left(z_1^{(\alpha\beta)} + p_1x_1\right)\right] - p_1\alpha_1[x_2 - h(x_1)] + k_1,\\
\text{for } (\alpha,\beta) &= (5,6),(7,8);\\
g_2^{(\alpha\beta)} &= z_1^{(\alpha\beta)} + p_1x_1 - \left(z_2^{(\alpha\beta)} + p_2x_2\right) + z_3^{(\alpha\beta)} + p_3x_3 - p_2(x_1 - x_2 + x_3) - k_2,\\
\text{for } (\alpha,\beta) &= (1,2),(5,6);\\
g_2^{(\alpha\beta)} &= z_1^{(\alpha\beta)} + p_1x_1 - \left(z_2^{(\alpha\beta)} + p_2x_2\right) + z_3^{(\alpha\beta)} + p_3x_3 - p_2(x_1 - x_2 + x_3) + k_2,\\
\text{for } (\alpha,\beta) &= (3,4),(7,8).
\end{aligned}
\right\} \tag{8.53}
$$

On the edge $\mathcal{E}_{l_1l_2l_1l_3}$, a dynamical system is

$$\begin{aligned}&\dot{\mathbf{z}}^{(l_1l_2l_1l_3)}=\mathbf{g}^{(l_1l_2l_1l_3)}\left(\mathbf{z}^{(l_1l_2l_1l_3)},\mathbf{x}(t),t\right);\\&\dot{\mathbf{x}}=\mathcal{F}(\mathbf{x},t)\end{aligned}\tag{8.54}$$

where

$$\mathbf{g}^{(l_1l_2l_1l_3)}=\left(g_1^{(l_1l_2l_1l_3)},g_2^{(l_1l_2l_1l_3)},g_3^{(l_1l_2l_1l_3)}\right)^{\mathrm{T}}.\tag{8.55}$$

Only the edge flow on the edge is presented. For a sliding flow on the edge $\mathcal{E}_{l_1l_2l_1l_3}$, we have

$$\left.\begin{aligned}&g_1^{(l_1l_2l_1l_3)}=z_1^{(l_1l_2l_1l_3)}=0\ \text{ on }\partial\Omega_{l_1l_2}\\&\text{for }\ (l_1,l_2)=\{(1,5),(2,6),(3,7),(4,8)\};\\&g_2^{(l_1l_2l_1l_3)}=z_2^{(l_1l_2l_1l_3)}=0\ \text{ on }\partial\Omega_{l_1l_2}\\&\text{for }\ (l_1,l_3)=\{(2,3),(1,4),(6,7),(8,5)\};\\&g_3^{(l_1l_2l_1l_3)}\equiv g_3^{(l_1)}=-\beta_2\left(z_2^{(l_1)}+p_2x_2\right)+p_3(\beta_1x_2)-k_3,\text{ for }l_1=1,4,5,8;\\&g_3^{(l_1l_2l_1l_3)}\equiv g_3^{(l_1)}=\beta_2\left(z_2^{(l_1)}+p_2x_2\right)+p_3(\beta_1x_2)+k_3,\ \ \text{for }l_1=2,3,6,7.\end{aligned}\right\}\tag{8.56}$$

$$\left.\begin{aligned}&g_1^{(l_1l_2l_1l_3)}=z_1^{(l_1l_2l_1l_3)}-p_1x_1=0\ \text{ on }\partial\Omega_{l_1l_2}\\&\text{for }\ (l_1,l_2)=\{(1,5),(2,6),(3,7),(4,8)\};\\&g_3^{(l_1l_2l_1l_3)}=z_3^{(l_1l_2l_1l_3)}-p_3x_3=0\ \text{ on }\partial\Omega_{l_1l_3}\\&\text{for }\ (l_1,l_3)=\{(1,2),(3,4),(5,6),(7,8)\}.\\&g_2^{(l_1l_2l_1l_3)}\equiv g_2^{(l_1)}(\mathbf{y}^{(l_1)},t)\\&=z_1^{(l_1)}+p_1x_1-\left(z_2^{(l_1)}+p_2x_2\right)+z_3^{(l_1)}+p_3x_3-p_2(x_1-x_2+x_3)-k_2,\\&\text{for }l_1=1,2,5,6;\\&g_2^{(l_1l_2l_1l_3)}\equiv g_2^{(l_1)}(\mathbf{y}^{(l_1)},t)\\&=z_1^{(l_1)}+p_1x_1-\left(z_2^{(l_1)}+p_2x_2\right)+z_3^{(l_1)}+p_3x_3-p_2(x_1-x_2+x_3)+k_2,\\&\text{for }l_1=3,4,7,8;\end{aligned}\right\}\tag{8.57}$$

$$\left.\begin{array}{l} g_2^{(l_1 l_2 l_1 l_3)} = z_2^{(l_1 l_2 l_1 l_3)} - p_2 x_2 = 0 \text{ on } \partial\Omega_{l_1 l_2} \\ \text{for } (l_1, l_2) = \{(2,3),(1,4),(6,7),(8,5)\}; \\ g_3^{(l_1 l_2 l_1 l_3)} = z_3^{(l_1 l_2 l_1 l_3)} - p_3 x_3 = 0 \text{ on } \partial\Omega_{l_1 l_3} \\ \text{for } (l_1, l_3) = \{(1,2),(3,4),(5,6),(7,8)\}; \\ g_1^{(l_1 l_2 l_1 l_3)} \equiv g_1^{(l_1)}(\mathbf{y}^{(l_1)}, t) \\ = \alpha_2 \left[z_2^{(l_1)} + p_2 x_2 - h\left(z_1^{(l_1)} + p_1 x_1 \right) \right] - p_1 \alpha_1 [x_2 - h(x_1)] - k_1 \\ \text{for } l_1 = 1,2,3,4; \\ g_1^{(l_1 l_2 l_1 l_3)} \equiv g_1^{(l_1)}(\mathbf{y}^{(l_1)}, t) \\ = \alpha_2 \left[z_2^{(l_1)} + p_2 x_2 - h\left(z_1^{(l_1)} + p_1 x_1 \right) \right] - p_1 \alpha_1 [x_2 - h(x_1)] + k_1, \\ \text{for } l_1 = 5,6,7,8 \end{array}\right\} \tag{8.58}$$

On the vertex $\mathcal{E}_0$, dynamical system is

$$\begin{aligned} \dot{\mathbf{z}}^{(0)} &= \mathbf{g}^{(0)}(\mathbf{z}^{(0)}, \mathbf{x}(t), t); \\ \dot{\mathbf{x}} &= \mathcal{F}(\mathbf{x}, t) \end{aligned} \tag{8.59}$$

where

$$\begin{aligned} \mathbf{g}^{(0)} &= \left(g_1^{(0)}, g_2^{(0)}, g_3^{(0)} \right)^{\mathrm{T}}, \\ g_1^{(0)} &= z_1^{(0)} = 0, \\ g_2^{(0)} &= z_2^{(0)} = 0, \\ g_2^{(0)} &= z_3^{(0)} = 0. \end{aligned} \tag{8.60}$$

From such a description, the theory of discontinuous dynamical systems will be used to obtain the analytical conditions for the projective synchronization of the two dynamical systems.

8.4 Analytical Synchronization Dynamics

As in Min and Luo (2015), from the theory of discontinuous dynamical system in Luo (2009a, b), the analytical conditions for projective synchronization of the two Chua circuits are presented. Herein, the G-functions in the relative coordinates are introduced for $z_m \in \partial\Omega_{ij}$ at $t = t_m$ as

$$G_{\partial\Omega_{ij}}^{(\alpha)}(\mathbf{z}_m,\mathbf{x},t_{m\pm})=\mathbf{n}_{\partial\Omega_{ij}}^{\mathrm{T}}\cdot\left[\mathbf{g}^{(\alpha)}(\mathbf{z}_m,\mathbf{x},t_{m\pm})-\mathbf{g}^{(ij)}(\mathbf{z}_m,\mathbf{x},t_{m\pm})\right],\tag{8.61}$$

$$G_{\partial\Omega_{ij}}^{(1,\alpha)}(\mathbf{z}_m,\mathbf{x},t_{m\pm})=\mathbf{n}_{\partial\Omega_{ij}}^{\mathrm{T}}\cdot\left[D\mathbf{g}^{(\alpha)}(\mathbf{z}_m,\mathbf{x},t_{m\pm})-D\mathbf{g}^{(ij)}(\mathbf{z}_m,\mathbf{x},t_{m\pm})\right]\tag{8.62}$$

where $G_{\partial\Omega_{ij}}^{(\alpha)}(\mathbf{z}_m,\mathbf{x},t_{m\pm})$ and $G_{\partial\Omega_{ij}}^{(1,\alpha)}(\mathbf{z}_m,\mathbf{x},t_{m\pm})$ are the zero-order and first-order G-functions of the flow in Ω_α ($\alpha\in\{i,j\}$) at the boundary $\partial\Omega_{ij}$ ($i,j=1,2,\cdots,8$) and $t_{m\pm}=t_m\pm 0$. The normal vectors of the relative boundaries in Eqs. 8.35–8.37 are

$$\begin{aligned}
&\mathbf{n}_{\partial\Omega_{12}}=\mathbf{n}_{\partial\Omega_{34}}=\mathbf{n}_{\partial\Omega_{56}}=\mathbf{n}_{\partial\Omega_{78}}=(0,0,1)^{\mathrm{T}},\\
&\mathbf{n}_{\partial\Omega_{23}}=\mathbf{n}_{\partial\Omega_{41}}=\mathbf{n}_{\partial\Omega_{67}}=\mathbf{n}_{\partial\Omega_{85}}=(0,1,0)^{\mathrm{T}},\\
&\mathbf{n}_{\partial\Omega_{15}}=\mathbf{n}_{\partial\Omega_{26}}=\mathbf{n}_{\partial\Omega_{37}}=\mathbf{n}_{\partial\Omega_{48}}=(1,0,0)^{\mathrm{T}}.
\end{aligned}\tag{8.63}$$

From Eqs. 8.45–8.60, the corresponding G-functions at boundary $\partial\Omega_{ij}$ ($i,j=1,2,\cdots,8$) are for domain Ω_α ($\alpha\in\{1,2,\cdots,8\}$)

$$\begin{aligned}
&G_{\partial\Omega_{ij}}^{(\alpha)}(\mathbf{z}_m,\mathbf{x},t_{m\pm})=g_3^{(\alpha)}(\mathbf{z}_m,\mathbf{x},t_{m\pm})\ \text{ for } (i,j)=(1,2),(3,4),(5,6),(7,8);\\
&G_{\partial\Omega_{ij}}^{(\alpha)}(\mathbf{z}_m,\mathbf{x},t_{m\pm})=g_2^{(\alpha)}(\mathbf{z}_m,\mathbf{x},t_{m\pm})\ \text{ for } (i,j)=(2,3),(4,1),(6,7),(8,5);\\
&G_{\partial\Omega_{ij}}^{(\alpha)}(\mathbf{z}_m,\mathbf{x},t_{m\pm})=g_1^{(\alpha)}(\mathbf{z}_m,\mathbf{x},t_{m\pm})\ \text{ for } (i,j)=(1,5),(2,5),(3,7),(4,8);\\
&G_{\partial\Omega_{ij}}^{(1,\alpha)}(\mathbf{z}_m,\mathbf{x},t_{m\pm})=Dg_3^{(\alpha)}(\mathbf{z}_m,\mathbf{x},t_{m\pm})\ \text{for } (i,j)=(1,2),(3,4),(5,6),(7,8)\\
&G_{\partial\Omega_{ij}}^{(1,\alpha)}(\mathbf{z}_m,\mathbf{x},t_{m\pm})=Dg_2^{(\alpha)}(\mathbf{z}_m,\mathbf{x},t_{m\pm})\ \text{for } (i,j)=(2,3),(4,1),(6,7),(8,5);\\
&G_{\partial\Omega_{ij}}^{(1,\alpha)}(\mathbf{z}_m,\mathbf{x},t_{m\pm})=Dg_1^{(\alpha)}(\mathbf{z}_m,\mathbf{x},t_{m\pm})\ \text{for } (i,j)=(1,5),(2,5),(3,7),(4,8).
\end{aligned}\tag{8.64}$$

where

$$\begin{aligned}
Dg_1^{(\alpha)}\left(\mathbf{z}^{(\alpha)},\mathbf{x},t\right)&=\alpha_2\{z_1^{(\alpha)}+p_1x_1-\left(z_2^{(\alpha)}+p_2x_2\right)+z_3^{(\alpha)}+p_3x_3\\
&\quad-\left[a+3b\left(z_1^{(\alpha)}+p_1x_1\right)^2\right]\}-p_1\alpha_1\left[x_1-x_2+x_3-(a+3bx_1^2)\right],\\
Dg_2^{(\alpha)}\left(\mathbf{z}^{(\alpha)},\mathbf{x},t\right)&=\alpha_2\left[z_2^{(\alpha)}+p_2x_2-h\left(z_1^{(\alpha)}+p_1x_1\right)\right]-z_1^{(\alpha)}+p_1x_1-\left(z_2^{(\alpha)}+p_2x_2\right)\\
&\quad+z_3^{(\alpha)}+p_3x_3-\beta_1\left(z_2^{(\alpha)}+p_2x_2\right)-p_2[\alpha_1(x_2-h(x_1))]\\
&\quad+p_2(x_1-x_2+x_3+\beta_1x_2x_3),\\
Dg_3^{(\alpha)}\left(\mathbf{z}^{(\alpha)},\mathbf{x},t\right)&=-\beta_1\left[z_1^{(\alpha)}+p_1x_1-\left(z_2^{(\alpha)}+p_2x_2\right)+z_3^{(\alpha)}+p_3x_3\right]+p_3\beta_2[x_1-x_2+x_3].
\end{aligned}\tag{8.65}$$

The G-functions in domains Ω_α ($\alpha \in \{1, 2, \cdots, 8\}$) with respect to the boundary are

$$
\begin{aligned}
&G^{(\alpha)}_{\partial\Omega_{ij}}\left(\mathbf{z}^{(\alpha)}, \mathbf{x}, t\right) = g^{(\alpha)}_3\left(\mathbf{z}^{(\alpha)}, \mathbf{x}, t\right) \text{ for } (i,j) = (1,2), (3,4), (5,6), (7,8);\\
&G^{(\alpha)}_{\partial\Omega_{ij}}\left(\mathbf{z}^{(\alpha)}, \mathbf{x}, t\right) = g^{(\alpha)}_2\left(\mathbf{z}^{(\alpha)}, \mathbf{x}, t\right) \text{ for } (i,j) = (2,3), (4,1), (6,7), (8,5);\\
&G^{(\alpha)}_{\partial\Omega_{ij}}\left(\mathbf{z}^{(\alpha)}, \mathbf{x}, t\right) = g^{(\alpha)}_1\left(\mathbf{z}^{(\alpha)}, \mathbf{x}, t\right) \text{ for } (i,j) = (1,5), (2,5), (3,7), (4,8);\\
&G^{(1,\alpha)}_{\partial\Omega_{ij}}\left(\mathbf{z}^{(\alpha)}, \mathbf{x}, t\right) = Dg^{(\alpha)}_3\left(\mathbf{z}^{(\alpha)}, \mathbf{x}, t\right) \text{ for } (i,j) = (1,2), (3,4), (5,6), (7,8)\\
&G^{(1,\alpha)}_{\partial\Omega_{ij}}\left(\mathbf{z}^{(\alpha)}, \mathbf{x}, t\right) = Dg^{(\alpha)}_2\left(\mathbf{z}^{(\alpha)}, \mathbf{x}, t\right) \text{ for } (i,j) = (2,3), (4,1), (6,7), (8,5);\\
&G^{(1,\alpha)}_{\partial\Omega_{ij}}\left(\mathbf{z}^{(\alpha)}, \mathbf{x}, t\right) = Dg^{(\alpha)}_1\left(\mathbf{z}^{(\alpha)}, \mathbf{x}, t\right) \text{ for } (i,j) = (1,5), (2,5), (3,7), (4,8).
\end{aligned}
\tag{8.66}
$$

From the theory for discontinuous dynamical systems (Min and Luo 2015; Luo 2009a), the conditions for projective synchronization of the two Chua circuits with different motions at the vertex point are

$$
\left.
\begin{aligned}
&G^{(1)}_{\partial\Omega_{15}}(\mathbf{z}_m, \mathbf{x}, t_{m-}) = g^{(1)}_1(\mathbf{z}_m, \mathbf{x}, t_{m-}) < 0,\\
&G^{(1)}_{\partial\Omega_{41}}(\mathbf{z}_m, \mathbf{x}, t_{m-}) = g^{(1)}_2(\mathbf{z}_m, \mathbf{x}, t_{m-}) < 0,\\
&G^{(1)}_{\partial\Omega_{12}}(\mathbf{z}_m, \mathbf{x}, t_{m-}) = g^{(1)}_3(\mathbf{z}_m, \mathbf{x}, t_{m-}) < 0,
\end{aligned}
\right\} \text{for } \mathbf{z}_m
$$
$$
\in \partial\Omega_{15} \cap \partial\Omega_{41} \cap \partial\Omega_{12} \text{ on } \Omega_1; \tag{8.67}
$$

$$
\left.
\begin{aligned}
&G^{(2)}_{\partial\Omega_{26}}(\mathbf{z}_m, \mathbf{x}, t_{m-}) = g^{(2)}_1(\mathbf{z}_m, \mathbf{x}, t_{m-}) < 0,\\
&G^{(2)}_{\partial\Omega_{23}}(\mathbf{z}_m, \mathbf{x}, t_{m-}) = g^{(2)}_2(\mathbf{z}_m, \mathbf{x}, t_{m-}) < 0,\\
&G^{(2)}_{\partial\Omega_{12}}(\mathbf{z}_m, \mathbf{x}, t_{m-}) = g^{(2)}_3(\mathbf{z}_m, \mathbf{x}, t_{m-}) > 0,
\end{aligned}
\right\} \text{for } \mathbf{z}_m
$$
$$
\in \partial\Omega_{26} \cap \partial\Omega_{23} \cap \partial\Omega_{12} \text{ on } \Omega_2; \tag{8.68}
$$

$$
\left.
\begin{aligned}
&G^{(3)}_{\partial\Omega_{37}}(\mathbf{z}_m, \mathbf{x}, t_{m-}) = g^{(3)}_1(\mathbf{z}_m, \mathbf{x}, t_{m-}) < 0,\\
&G^{(3)}_{\partial\Omega_{23}}(\mathbf{z}_m, \mathbf{x}, t_{m-}) = g^{(3)}_2(\mathbf{z}_m, \mathbf{x}, t_{m-}) > 0,\\
&G^{(3)}_{\partial\Omega_{34}}(\mathbf{z}_m, \mathbf{x}, t_{m-}) = g^{(3)}_3(\mathbf{z}_m, \mathbf{x}, t_{m-}) > 0,
\end{aligned}
\right\} \text{for } \mathbf{z}_m
$$
$$
\in \partial\Omega_{37} \cap \partial\Omega_{23} \cap \partial\Omega_{34} \text{ on } \Omega_3; \tag{8.69}
$$

$$\left.\begin{array}{l} G^{(4)}_{\partial\Omega_{48}}(\mathbf{z}_m,\mathbf{x},t_{m-}) = g^{(4)}_1(\mathbf{z}_m,\mathbf{x},t_{m-}) < 0, \\ G^{(4)}_{\partial\Omega_{41}}(\mathbf{z}_m,\mathbf{x},t_{m-}) = g^{(4)}_2(\mathbf{z}_m,\mathbf{x},t_{m-}) > 0, \\ G^{(4)}_{\partial\Omega_{34}}(\mathbf{z}_m,\mathbf{x},t_{m-}) = g^{(4)}_3(\mathbf{z}_m,\mathbf{x},t_{m-}) < 0, \end{array}\right\} \text{for } \mathbf{z}_m$$
$$\in \partial\Omega_{48} \cap \partial\Omega_{41} \cap \partial\Omega_{34} \text{ on } \Omega_4; \tag{8.70}$$

$$\left.\begin{array}{l} G^{(5)}_{\partial\Omega_{15}}(\mathbf{z}_m,\mathbf{x},t_{m-}) = g^{(5)}_1(\mathbf{z}_m,\mathbf{x},t_{m-}) > 0, \\ G^{(5)}_{\partial\Omega_{85}}(\mathbf{z}_m,\mathbf{x},t_{m-}) = g^{(5)}_2(\mathbf{z}_m,\mathbf{x},t_{m-}) < 0, \\ G^{(5)}_{\partial\Omega_{56}}(\mathbf{z}_m,\mathbf{x},t_{m-}) = g^{(5)}_3(\mathbf{z}_m,\mathbf{x},t_{m-}) < 0, \end{array}\right\} \text{for } \mathbf{z}_m$$
$$\in \partial\Omega_{15} \cap \partial\Omega_{85} \cap \partial\Omega_{56} \text{ on } \Omega_5; \tag{8.71}$$

$$\left.\begin{array}{l} G^{(6)}_{\partial\Omega_{26}}(\mathbf{z}_m,\mathbf{x},t_{m-}) = g^{(6)}_1(\mathbf{z}_m,\mathbf{x},t_{m-}) > 0, \\ G^{(6)}_{\partial\Omega_{67}}(\mathbf{z}_m,\mathbf{x},t_{m-}) = g^{(6)}_2(\mathbf{z}_m,\mathbf{x},t_{m-}) < 0, \\ G^{(6)}_{\partial\Omega_{56}}(\mathbf{z}_m,\mathbf{x},t_{m-}) = g^{(6)}_3(\mathbf{z}_m,\mathbf{x},t_{m-}) > 0, \end{array}\right\} \text{for } \mathbf{z}_m$$
$$\in \partial\Omega_{26} \cap \partial\Omega_{67} \cap \partial\Omega_{56} \text{ on } \Omega_6; \tag{8.72}$$

$$\left.\begin{array}{l} G^{(7)}_{\partial\Omega_{37}}(\mathbf{z}_m,\mathbf{x},t_{m-}) = g^{(7)}_1(\mathbf{z}_m,\mathbf{x},t_{m-}) > 0, \\ G^{(7)}_{\partial\Omega_{67}}(\mathbf{z}_m,\mathbf{x},t_{m-}) = g^{(7)}_2(\mathbf{z}_m,\mathbf{x},t_{m-}) > 0, \\ G^{(7)}_{\partial\Omega_{78}}(\mathbf{z}_m,\mathbf{x},t_{m-}) = g^{(7)}_3(\mathbf{z}_m,\mathbf{x},t_{m-}) > 0, \end{array}\right\} \text{for } \mathbf{z}_m$$
$$\in \partial\Omega_{37} \cap \partial\Omega_{67} \cap \partial\Omega_{78} \text{ on } \Omega_7; \tag{8.73}$$

$$\left.\begin{array}{l} G^{(8)}_{\partial\Omega_{48}}(\mathbf{z}_m,\mathbf{x},t_{m-}) = g^{(8)}_1(\mathbf{z}_m,\mathbf{x},t_{m-}) > 0, \\ G^{(8)}_{\partial\Omega_{58}}(\mathbf{z}_m,\mathbf{x},t_{m-}) = g^{(8)}_2(\mathbf{z}_m,\mathbf{x},t_{m-}) > 0, \\ G^{(8)}_{\partial\Omega_{78}}(\mathbf{z}_m,\mathbf{x},t_{m-}) = g^{(8)}_3(\mathbf{z}_m,\mathbf{x},t_{m-}) < 0, \end{array}\right\} \text{for } \mathbf{z}_m$$
$$\in \partial\Omega_{48} \cap \partial\Omega_{58} \cap \partial\Omega_{78} \text{ on } \Omega_8. \tag{8.74}$$

From Eq. 8.66, six basic functions are defined as

$$
\begin{aligned}
&g_1\left(\mathbf{z}^{(\alpha)},\mathbf{x},t\right) \equiv g_1^{(\alpha)}\left(\mathbf{z}^{(\alpha)},\mathbf{x},t\right)\\
&=\alpha_2\left[z_2^{(\alpha)}+p_2x_2-h\left(z_1^{(\alpha)}+p_1x_1\right)\right]-p_1\alpha_1[x_2-h(x_1)]-k_1,\\
&\text{for } \alpha=1,2,3,4;\\
&g_2\left(\mathbf{z}^{(\alpha)},\mathbf{x},t\right) \equiv g_1^{(\alpha)}\left(\mathbf{z}^{(\alpha)},\mathbf{x},t\right)\\
&=\alpha_2\left(z_2^{(\alpha)}+p_2x_2-h\left(z_1^{(\alpha)}+p_1x_1\right)\right)-p_1\alpha_1(x_2-h(x_1))+k_1,\\
&\text{for } \alpha=5,6,7,8;\\
&g_3\left(\mathbf{z}^{(\alpha)},\mathbf{x},t\right) \equiv g_2^{(\alpha)}\left(\mathbf{z}^{(\alpha)},\mathbf{x},t\right)\\
&=z_1^{(\alpha)}+p_1x_1-\left(z_2^{(\alpha)}+p_2x_2\right)+z_3^{(\alpha)}+p_3x_3-p_2(x_1-x_2+x_3)-k_2,\\
&\text{for } \alpha=1,2,5,6;\\
&g_4\left(\mathbf{z}^{(\alpha)},\mathbf{x},t\right) \equiv g_2^{(\alpha)}\left(\mathbf{z}^{(\alpha)},\mathbf{x},t\right)\\
&=z_1^{(\alpha)}+p_1x_1-\left(z_2^{(\alpha)}+p_2x_2\right)+z_3^{(\alpha)}+p_3x_3-p_2(x_1-x_2+x_3)+k_2,\\
&\text{for } \alpha=3,4,7,8;\\
&g_5\left(\mathbf{z}^{(\alpha)},\mathbf{x},t\right) \equiv g_3^{(\alpha)}\left(\mathbf{z}^{(\alpha)},\mathbf{x},t\right)=-\beta_1\left(z_2^{(\alpha)}+p_2x_2\right)-p_3(-\beta_2x_2)-k_3,\\
&\text{for } \alpha=1,4,5,8;\\
&g_6\left(\mathbf{z}^{(\alpha)},\mathbf{x},t\right) \equiv g_3^{(\alpha)}\left(\mathbf{z}^{(\alpha)},\mathbf{x},t\right)=-\beta_1\left(z_2^{(\alpha)}+p_2x_2\right)-p_3(-\beta_2x_2)+k_3\\
&\text{for } \alpha=2,3,6,7.
\end{aligned}
\tag{8.75}
$$

Then the projective synchronization condition in Eqs. 8.67–8.74 becomes

$$
\begin{aligned}
g_1(\mathbf{z}_m,\mathbf{x},t_{m-}) &=\alpha_2[z_{2m}+p_2x_2-h(z_{1m}+p_1x_1))-p_1\alpha_1(x_2-h(x_1)]-k_1<0,\\
g_2(\mathbf{z}_m,\mathbf{x},t_{m-}) &=\alpha_2(z_{2m}+p_2x_2-h(z_{1m}+p_1x_1))-p_1\alpha_1(x_2-h(x_1))+k_1>0,\\
g_3(\mathbf{z}_m,\mathbf{x},t_{m-}) &=z_{1m}+p_1x_1-(z_{2m}+p_2x_2)+z_{3m}+p_3x_3-p_2(x_1-x_2+x_3)-k_2<0,\\
g_4(\mathbf{z}_m,\mathbf{x},t_{m-}) &=z_{1m}+p_1x_1-(z_{2m}+p_2x_2)+z_{3m}+p_3x_3-p_2(x_1-x_2+x_3)+k_2>0,\\
g_5(\mathbf{z}_m,\mathbf{x},t_{m-}) &=-\beta_1(z_{2m}+p_2x_2)-p_3(-\beta_2x_2)-k_3<0,\\
g_6(\mathbf{z}_m,\mathbf{x},t_{m-}) &=-\beta_1(z_{2m}+p_2x_2)-p_3(-\beta_2x_2)+k_3>0.
\end{aligned}
\tag{8.76}
$$

From Eq. 8.76, setting $\mathbf{z}_m=\mathbf{0}$, the conditions of projective synchronization for the controlled Chua system with the expected chaotic Chua circuit are

$$\begin{aligned}
g_1(\mathbf{z}_m,\mathbf{x},t_{m-}) &= \alpha_2[p_2x_2 - h(p_1x_1)] - p_1\alpha_1[x_2 - h(x_1)] - k_1 < 0,\\
g_2(\mathbf{z}_m,\mathbf{x},t_{m-}) &= \alpha_2[p_2x_2 - h(p_1x_1)] - p_1\alpha_1[x_2 - h(x_1)] + k_1 > 0,\\
g_3(\mathbf{z}_m,\mathbf{x},t_{m-}) &= p_1x_1 - p_2x_2 + p_3x_3 - p_2(x_1 - x_2 + x_3) - k_2 < 0,\\
g_4(\mathbf{z}_m,\mathbf{x},t_{m-}) &= p_1x_1 - p_2x_2 + p_3x_3 - p_2(x_1 - x_2 + x_3) + k_2 > 0,\\
g_5(\mathbf{z}_m,\mathbf{x},t_{m-}) &= -\beta_1p_2x_2 + \beta_2p_3x_2 - k_3 < 0,\\
g_6(\mathbf{z}_m,\mathbf{x},t_{m-}) &= -\beta_1p_2x_2 + p_3\beta_2x_2 + k_3 > 0
\end{aligned} \tag{8.77}$$

where

$$h(p_1x_1) = ap_1x_1 + b(p_1x_1)^3 \tag{8.78}$$

and the synchronization invariant domain is determined by

$$\begin{aligned}
\alpha_2p_2x_2 - \alpha_2h(p_1x_1) - p_1\alpha_1[x_2 - h(x_1)] &< k_1 < p_1\alpha_1[x_2 - h(x_1)] - \alpha_2p_2x_2 + \alpha_2h(p_1x_1),\\
p_1x_1 - p_2x_2 + p_3x_3 - p_2(x_1 - x_2 + x_3) &< k_2 < p_2(x_1 - x_2 + x_3) - p_1x_1 + p_2x_2 - p_3x_3,\\
-\beta_1p_2x_2 + p_3\beta_2x_2 &< k_3 < \beta_1p_2x_2 - p_3\beta_2x_2.
\end{aligned} \tag{8.79}$$

From Luo (Min and Luo 2015; Luo 2009a), the vanishing conditions of projective synchronization for two coupled systems with $\mathbf{z}_m^{(\alpha)} = \mathbf{z}_m$ on $\partial\Omega_{\alpha\beta}$ for $(\alpha,\beta) = (1,5)$, $(2,6)$, $(3,7)$, $(4,8)$ are

$$\left.\begin{aligned}
g_1(\mathbf{z}_m^{(\alpha)},\mathbf{x},t_{m\mp}) &= \alpha_2\left[z_{2m}^{(\alpha)} + p_2x_2 - h\left(z_{1m}^{(\alpha)} + p_1x_1\right)\right] - p_1\alpha_1[x_2 - h(x_1)] - k_1 = 0,\\
, Dg_1(\mathbf{z}_m^{(\alpha)},\mathbf{x},t_{m\mp}) &= \Gamma_1(\mathbf{z}_m^{(\alpha)},\mathbf{x},t_{m\mp}) > 0,\\
g_2(\mathbf{z}_m^{(\beta)},\mathbf{x},t_{m-}) &= \alpha_2\left[z_{2m}^{(\beta)} + p_2x_2 - h\left(z_{1m}^{(\beta)} + p_1x_1\right)\right] - p_1\alpha_1[x_2 - h(x_1)] + k_1 > 0;
\end{aligned}\right\} \tag{8.80}$$

for $z_{m+\varepsilon} = y_1 - p_1x_1 > 0$ and

$$\left.\begin{aligned}
g_1(\mathbf{z}_m^{(\alpha)},\mathbf{x},t_{m\mp}) &= \alpha_2\left[z_{2m}^{(\alpha)} + p_2x_2 - h\left(z_{1m}^{(\alpha)} + p_1x_1\right)\right] - p_1\alpha_1[x_2 - h(x_1)] - k_1 < 0,\\
g_2(\mathbf{z}_m^{(\beta)},\mathbf{x},t_{m-}) &= \alpha_2\left[z_{2m}^{(\beta)} + p_2x_2 - h\left(z_{1m}^{(\beta)} + p_1x_1\right)\right] - p_1\alpha_1[x_2 - h(x_1)] + k_1 = 0,\\
Dg_2(\mathbf{z}_m^{(\beta)},\mathbf{x},t_{m\mp}) &= \Gamma_1(\mathbf{z}_m^{(\beta)},\mathbf{x},t_{m\mp}) < 0;
\end{aligned}\right\} \tag{8.81}$$

for $z_{m+\varepsilon} = y_1 - p_1x_1 < 0$ with

$$\begin{aligned}\Gamma_1\left(\mathbf{z}_m^{(\alpha)},\mathbf{x},t_{m\mp}\right) &= \alpha_2\{z_{1m}^{(\alpha)}+p_1x_1-\left(z_{2m}^{(\alpha)}+p_2x_2\right)+z_{3m}^{(\alpha)}+p_3x_3\\ &\quad -\left[a+3b\left(z_{1m}^{(\alpha)}+p_1x_1\right)^2\right]\} - p_1\alpha_1\left[x_1-x_2+x_3-(a+3bx_1^2)\right],\end{aligned} \tag{8.82}$$

for $\alpha \in \{1,2,\cdots,8\}$—$\alpha$ can be replaced by β.

The vanishing conditions of projective synchronization on second boundary $\partial\Omega_{\alpha\beta}$ for $(\alpha,\beta) = (1,4), (2,3), (5,8), (6,7)$ are

$$\left.\begin{aligned} g_3\left(\mathbf{z}_m^{(\alpha)},\mathbf{x},t_{m\mp}\right) &= z_{1m}^{\alpha}+p_1x_1-\left(z_{2m}^{\alpha}+p_2x_2\right)+z_{3m}^{\alpha}+p_3x_3-p_2(x_1-x_2+x_3)-k_2=0,\\ Dg_3\left(\mathbf{z}_m^{(\alpha)},\mathbf{x},t_{m\mp}\right) &= \Gamma_2\left(\mathbf{z}_m^{(\alpha)},\mathbf{x},t_{m\mp}\right)>0,\\ g_4\left(\mathbf{z}_m^{(\beta)},\mathbf{x},t_{m-}\right) &= z_{1m}^{\alpha}+p_1x_1-\left(z_{2m}^{\alpha}+p_2x_2\right)+z_{3m}^{\alpha}+p_3x_3-p_2(x_1-x_2+x_3)+k_2>0; \end{aligned}\right\} \tag{8.83}$$

for $\dot{z}_{m+\varepsilon} = y_2 - p_2x_2 > 0$, and

$$\left.\begin{aligned} g_3\left(\mathbf{z}_m^{(\alpha)},\mathbf{x},t_{m-}\right) &= z_{1m}^{\alpha}+p_1x_1-\left(z_{2m}^{\alpha}+p_2x_2\right)+z_{3m}^{\alpha}+p_3x_3-p_2(x_1-x_2+x_3)-k_2<0,\\ g_4\left(\mathbf{z}_m^{(\beta)},\mathbf{x},t_{m\mp}\right) &= z_{1m}^{\alpha}+p_1x_1-\left(z_{2m}^{\alpha}+p_2x_2\right)+z_{3m}^{\alpha}+p_3x_3-p_2(x_1-x_2+x_3)+k_2=0,\\ Dg_4\left(\mathbf{z}_m^{(\beta)},\mathbf{x},t_{m\mp}\right) &= \Gamma_2\left(\mathbf{z}_m^{(\beta)},\mathbf{x},t_{m\mp}\right)<0; \end{aligned}\right\} \tag{8.84}$$

for $\dot{z}_{m+\varepsilon} = y_2 - p_2x_2 < 0$ with

$$\begin{aligned}\Gamma_2\left(\mathbf{z}_m^{(\alpha)},\mathbf{x},t_{m\mp}\right) &= \alpha_2\left(z_{2m}^{(\alpha)}+p_2x_2-h\left(z_{1m}^{(\alpha)}+p_1x_1\right)\right)-z_{1m}^{(\alpha)}+p_1x_1-\left(z_{2m}^{(\alpha)}+p_2x_2\right)\\ &\quad +z_{3m}^{(\alpha)}+p_3x_3-\beta_1\left(z_{2m}^{(\alpha)}+p_2x_2\right)-p_2(\alpha_1(x_2-h(x_1)))\\ &\quad +p_2(x_1-x_2+x_3+\beta_1x_2x_3).\end{aligned} \tag{8.85}$$

for $\alpha \in \{1,2,3,4,5,6,7,8\}$—and α can be replaced by β.

The vanishing conditions of projective synchronization on the boundary $\partial\Omega_{\alpha\beta}$ for $(\alpha,\beta) = \{(1,2),(3,4),(5,6),(7,8)\}$ are

$$\left.\begin{aligned} g_5\left(\mathbf{z}_m^{(\alpha)},\mathbf{x},t_{m\mp}\right) &= -\beta_1\left(z_{2m}^{\alpha}+p_2x_2\right)-p_3(-\beta_2x_2)-k_3=0,\\ ,\ Dg_5\left(\mathbf{z}_m^{(\alpha)},\mathbf{x},t_{m\mp}\right) &= \Gamma_3\left(\mathbf{z}_m^{(\alpha)},\mathbf{x},t_{m\mp}\right)>0,\\ g_6\left(\mathbf{z}_m^{(\beta)},\mathbf{x},t_{m-}\right) &= -\beta_1\left(z_{2m}^{\alpha}+p_2x_2\right)-p_3(-\beta_2x_2)+k_3>0; \end{aligned}\right\} \tag{8.86}$$

for $\dot{z}_{m+\varepsilon} = y_3 - p_3x_3 > 0$, and

$$\left.\begin{aligned} g_5\left(\mathbf{z}_m^{(\alpha)},\mathbf{x},t_{m-}\right) &= -\beta_1\left(z_{2m}^{\alpha}+p_2x_2\right)-p_3(-\beta_2x_2)-k_3<0,\\ g_6\left(\mathbf{z}_m^{(\beta)},\mathbf{x},t_{m\mp}\right) &= -\beta_1\left(z_{2m}^{\alpha}+p_2x_2\right)-p_3(-\beta_2x_2)+k_3=0,\\ Dg_6\left(\mathbf{z}_m^{(\beta)},\mathbf{x},t_{m\mp}\right) &=\Gamma_3\left(\mathbf{z}_m^{(\beta)},\mathbf{x},t_{m\mp}\right)<0; \end{aligned}\right\} \tag{8.87}$$

For $\dot{z}_{m+\varepsilon}=y_3-p_3x_3<0,$ with

$$\begin{aligned} \Gamma_3\left(\mathbf{z}_m^{(\alpha)},\mathbf{x},t_{m\mp}\right) = -\beta_1\left[z_{1m}^{(\alpha)}+p_1x_1-\left(z_{2m}^{(\alpha)}+p_2x_2\right)+z_{3m}^{(\alpha)}+p_3x_3\right]\\ +p_3\beta_2(x_1-x_2+x_3) \end{aligned} \tag{8.88}$$

for $\alpha\in\{1,2,\cdots,8\}$—and α can be replaced by β.

The onset conditions of projective synchronization for the controlled slave system on the first boundary $\partial\Omega_{\alpha\beta}$ for $(\alpha,\beta)=\{(1,2),(3,4),(5,6),(7,8)\}$ are

$$\left.\begin{aligned} g_1\left(\mathbf{z}_m^{(\alpha)},\mathbf{x},t_{m\pm}\right) &=\alpha_2\left(z_{2m}^{(\alpha)}+p_2x_2-h\left(z_{1m}^{(\alpha)}+p_1x_1\right)\right)-p_1\alpha_1(x_2-h(x_1))-k_1=0,\\ ,\ Dg_1\left(\mathbf{z}_m^{(\alpha)},\mathbf{x},t_{m\pm}\right) &=\Gamma_1\left(\mathbf{z}_m^{(\alpha)},\mathbf{x},t_{m\pm}\right)>0,\\ g_2\left(\mathbf{z}_m^{(\beta)},\mathbf{x},t_{m-}\right) &=\alpha_2\left(z_{2m}^{(\beta)}+p_2x_2-h\left(z_{1m}^{(\beta)}+p_1x_1\right)\right)-p_1\alpha_1(x_2-h(x_1))+k_1>0; \end{aligned}\right\} \tag{8.89}$$

for $z_{m+\varepsilon}=y_1-p_1x_1>0.$

$$\left.\begin{aligned} g_1\left(\mathbf{z}_m^{(\alpha)},\mathbf{x},t_{m\pm}\right) &=\alpha_2\left(z_{2m}^{(\alpha)}+p_2x_2-h\left(z_{1m}^{(\alpha)}+p_1x_1\right)\right)-p_1\alpha_1(x_2-h(x_1))-k_1<0,\\ g_2\left(\mathbf{z}_m^{(\beta)},\mathbf{x},t_{m-}\right) &=\alpha_2\left(z_{2m}^{(\beta)}+p_2x_2-h\left(z_{1m}^{(\beta)}+p_1x_1\right)\right)-p_1\alpha_1(x_2-h(x_1))+k_1=0,\\ Dg_2\left(\mathbf{z}_m^{(\beta)},\mathbf{x},t_{m\pm}\right) &=\Gamma_1\left(\mathbf{z}_m^{(\beta)},\mathbf{x},t_{m\pm}\right)<0; \end{aligned}\right\} \tag{8.90}$$

for $z_{m+\varepsilon}=y_1-p_1x_1<0.$

The onset conditions of projective synchronization for the controlled slave system on the second boundary $\partial\Omega_{\alpha\beta}$ for $(\alpha,\beta)=\{(1,2),(3,4),(5,6),(7,8)\}$ are

$$\left.\begin{aligned} g_3\left(\mathbf{z}_m^{(\alpha)},\mathbf{x},t_{m\pm}\right) &=z_{1m}^{\alpha}+p_1x_1-\left(z_{2m}^{\alpha}+p_2x_2\right)+z_{3m}^{\alpha}+p_3x_3-p_2(x_1-x_2+x_3)-k_2=0,\\ Dg_3\left(\mathbf{z}_m^{(\alpha)},\mathbf{x},t_{m\pm}\right) &=\Gamma_2\left(\mathbf{z}_m^{(\alpha)},\mathbf{x},t_{m\pm}\right)>0,\\ g_4\left(\mathbf{z}_m^{(\beta)},\mathbf{x},t_{m-}\right) &=z_{1m}^{\alpha}+p_1x_1-\left(z_{2m}^{\alpha}+p_2x_2\right)+z_{3m}^{\alpha}+p_3x_3-p_2(x_1-x_2+x_3)+k_2>0; \end{aligned}\right\} \tag{8.91}$$

for $\dot{z}_{m+\varepsilon}=y_2-p_2x_2>0,$ and

$$\left.\begin{aligned} g_3\left(\mathbf{z}_m^{(\alpha)},\mathbf{x},t_{m-}\right) &= z_{1m}^{\alpha}+p_1x_1-\left(z_{2m}^{\alpha}+p_2x_2\right)+z_{3m}^{\alpha}+p_3x_3-p_2(x_1-x_2+x_3)-k_2<0,\\ g_4\left(\mathbf{z}_m^{(\beta)},\mathbf{x},t_{m\pm}\right) &= z_{1m}^{\alpha}+p_1x_1-\left(z_{2m}^{\alpha}+p_2x_2\right)+z_{3m}^{\alpha}+p_3x_3-p_2(x_1-x_2+x_3)+k_2=0,\\ Dg_4\left(\mathbf{z}_m^{(\beta)},\mathbf{x},t_{m\pm}\right) &= \Gamma_2\left(\mathbf{z}_m^{(\beta)},\mathbf{x},t_{m\pm}\right)<0; \end{aligned}\right\} \tag{8.92}$$

for $\dot{z}_{m+\varepsilon}=y_2-p_2x_2<0$.

The onset conditions of projective synchronization for the controlled slave system on the third boundary $\partial\Omega_{\alpha\beta}$ for $(\alpha,\beta)=\{(1,2),(3,4),(5,6),(7,8)\}$ are

$$\left.\begin{aligned} g_5\left(\mathbf{z}_m^{(\alpha)},\mathbf{x},t_{m\pm}\right) &= -\beta_1\left(z_{2m}^{\alpha}+p_2x_2\right)-p_3(-\beta_2x_2)-k_3=0,\\ Dg_5\left(\mathbf{z}_m^{(\alpha)},\mathbf{x},t_{m\pm}\right) &= \Gamma_3\left(\mathbf{z}_m^{(\alpha)},\mathbf{x},t_{m\pm}\right)>0,\\ g_6\left(\mathbf{z}_m^{(\beta)},\mathbf{x},t_{m-}\right) &= -\beta_1\left(z_{2m}^{\alpha}+p_2x_2\right)-p_3(-\beta_2x_2)+k_3>0; \end{aligned}\right\} \tag{8.93}$$

for $\dot{z}_{m+\varepsilon}=y_3-p_3x_3>0$, and

$$\left.\begin{aligned} g_5\left(\mathbf{z}_m^{(\alpha)},\mathbf{x},t_{m-}\right) &= -\beta_1\left(z_{2m}^{\alpha}+p_2x_2\right)-p_3(-\beta_2x_2)-k_3<0,\\ g_6\left(\mathbf{z}_m^{(\beta)},\mathbf{x},t_{m\pm}\right) &= -\beta_1\left(z_{2m}^{\alpha}+p_2x_2\right)-p_3(-\beta_2x_2)+k_3=0,\\ Dg_6\left(\mathbf{z}_m^{(\beta)},\mathbf{x},t_{m\pm}\right) &= \Gamma_3\left(\mathbf{z}_m^{(\beta)},\mathbf{x},t_{m\pm}\right)<0; \end{aligned}\right\} \tag{8.94}$$

for $\dot{z}_{m+\varepsilon}=y_3-p_3x_3<0$.

8.5 Illustrations of Synchronization Dynamics

As in Min and Luo (2015), consider a set of parameters and initial conditions

$$\begin{aligned} &a=-1/7, b=2/7, \alpha_1=10.0068, \alpha_2=9.01415,\\ &\beta_1=\beta_2=100/7, p_1=0.5, p_2=2.0, p_3=1.5,\\ &(x_1,x_2,x_3)=(0.74195, 0.01496, -0.74526), \text{and}\\ &(y_1,y_2,y_3)=(0.31556, 0.083599, -0.028979). \end{aligned} \tag{8.95}$$

The master system possesses the double scroll chaotic attractor, and the slave system without control law experiences a single scroll chaotic motion with the above parameters and initial conditions. The synchronization scenarios varying with parameters and the parameter maps are developed, and numerical illustrations of partial and full synchronization will be presented.

8.5.1 Synchronization Scenarios

As in Min and Luo (2015), from the analytical conditions in Eqs. 8.67–8.94, the projective synchronization scenario of two coupled Chua circuits with different motions is presented in Fig. 8.2. Switching points varying with the control parameters are shown through ($y_{1k} = p_1 x_{1k}$, $y_{2k} = p_2 x_{2k}$ and $y_{3k} = p_3 x_{3k}$). The acronyms "NS," "FS," and "PS" mean "non-synchronization," "full synchronization," and "partial synchronization," respectively. From the analytical conditions for synchronization, the switching points varying with the control parameter k_1 are plotted in Fig. 8.2a–c with $k_2 = k_3 = 2.0$. The partial projective synchronization of two Chua circuits exists in the range of $k_1 \in (0.023, 2.21)$. For $k_1 \in (0, 0.023)$, the projective synchronization for two chaotic motions cannot be obtained. For $k_1 \in (2.21, \infty)$, the full synchronization of chaotic motions for the two Chua systems is achieved.

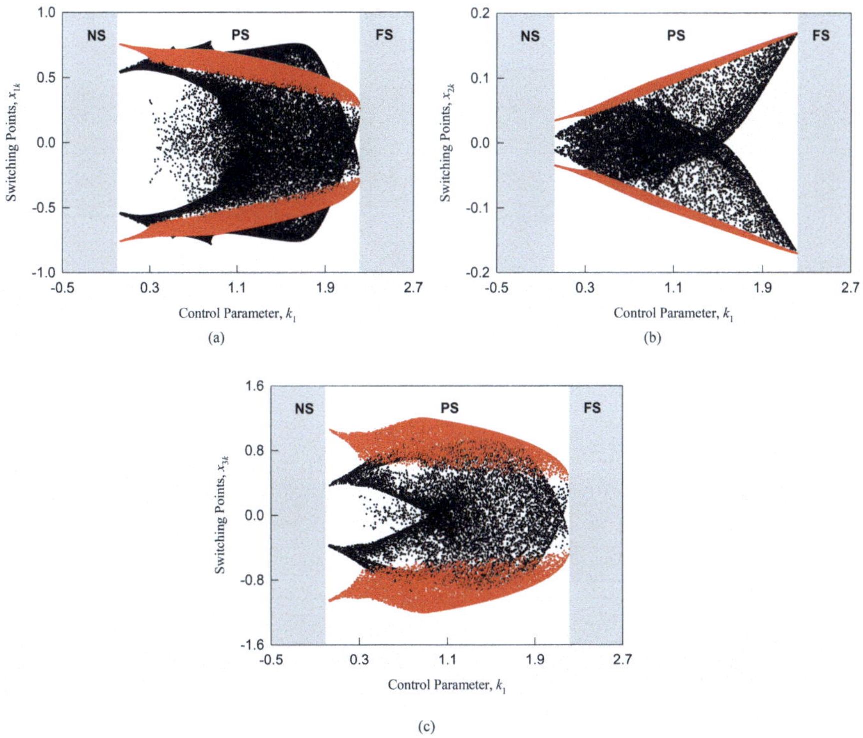

Fig. 8.2 The projective synchronization scenario of two Chua systems. (**a**–**c**) Switching points $x_i (i = 1, 2, 3)$ varying with k_1 for $k_2 = k_3 = 2.0$. (*FS* full synchronization, *PS* partial synchronization, *NS* non-synchronization, *black points* synchronization appearance, *red points* synchronization disappearance)

8.5.2 Parameter Maps

As in Min and Luo (2015), for a global view of projective function synchronization, the parameter maps are presented in Fig. 8.3. Non-synchronization, partial synchronization, and full synchronization are labeled by the corresponding acronyms "NS," "PS," and "FS," respectively. In Fig. 8.3a, the control parameter map (k_1, k_2) with $k_3 = 2.0$ is plotted. For $k_1 > 2.21$ and $k_2 > 0.84$, there exists the projective synchronization of the two systems. However, only the partial synchronization exists in the ranges of $0.0 < k_1 < 0.84$ and $k_1 > 2.21$. For small k_1, the non-synchronization area is observed. In Fig. 8.3b, the control parameter map (k_2, k_3) with $k_1 = 3.0$ is presented. The full synchronization of the two Chua systems is for $k_2 > 0.84$ and $k_3 > 1.214$, and the non-synchronization area is observed for small k_2. The control parameter map (k_1, k_3) with $k_2 = 2.0$ is also developed, as shown in Fig. 8.3c. The full synchronization of the two Chua systems lies in $k_1 > 2.21$ and $k_3 > 1.214$ at $k_2 = 2.0$. The parameter map (α_1, k_1) with $k_2 = k_3 = 2.0$ is also generated from the

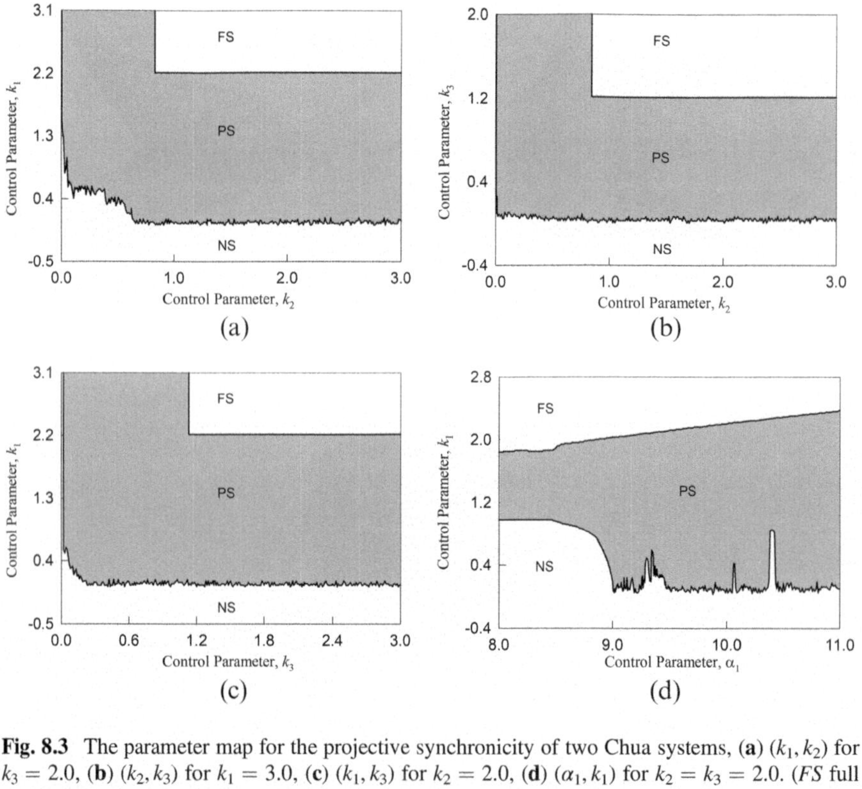

Fig. 8.3 The parameter map for the projective synchronicity of two Chua systems, (**a**) (k_1, k_2) for $k_3 = 2.0$, (**b**) (k_2, k_3) for $k_1 = 3.0$, (**c**) (k_1, k_3) for $k_2 = 2.0$, (**d**) (α_1, k_1) for $k_2 = k_3 = 2.0$. (*FS* full synchronization, *PS*: partial synchronization, *NS* non-synchronization)

analytical conditions, as presented in Fig. 8.3d. As the parameter α_1 increases, the parameter region of partial synchronization becomes larger, and the control parameter k_1 becomes large as well. With varying α_1, the boundary portion of the partial synchronization is not smooth due to the chaotic motion, but the bottom boundary smooth portion is owing to the periodic motion.

8.5.3 Synchronization Illustrations

As in Min and Luo (2015), from the parameter maps, the control parameters k_i $(i = 1, 2, 3)$ can be chosen for the numerical illustration of synchronizations of the two Chua systems. The partial projective synchronization of the two Chua systems can be obtained, as shown in Figs. 8.4, 8.5 and 8.6 with different control parameters. The symbols "S" and "N" denote "synchronization" and "non-synchronization," respectively. Hollow white circular symbols represent synchronization appearance, and filled green circular symbols means synchronization disappearance. The big gray circular symbols are initial conditions.

From the parameter map of (k_1, k_2) with $k_3 = 2.0$, the partial projective synchronization of two Chua systems at $k_1 = 1.5, k_2 = 2.0, k_3 = 2.0$ is plotted in Fig. 8.5 for a small parameter k_1. The trajectories for the master Chua system are depicted through (x_1, x_2) and (x_1, x_3) planes in Fig. 8.4a, b. The synchronization domain of the master system with $(x_3 = 0)$ is superimposed in the (x_1, x_2)plane. The white and shaded areas are for synchronization and non-synchronization domains. In Fig. 8.4a, it is clearly observed that the synchronization vanishing exists on the boundaries between the synchronization and non-synchronization domains. The synchronization appearance is in the synchronization domain with the white hollow symbols. In Fig. 8.4c–e, all G-functions varying with state variable x_1 are presented. In Fig. 8.4c, the G-functions $g_i(i = 1, 2)$ for the synchronization of state variables x_1 and y_1 are presented. From the synchronization conditions, if $g_1 < 0$ with $g_2 > 0$, the two state variables x_1 and y_1 will be synchronized; if $g_1 < 0$ with $g_2 > 0$, the two state variables x_1 and y_1 cannot be synchronized. It is observed that the two state variables x_1 and y_1 are partially synchronized. In Fig. 8.4d, the G-functions of $g_i(i = 3, 4)$ for the synchronization of state variables x_2 and y_2 are presented. Because of $g_3 < 0$ and $g_4 > 0$, the two state variables x_2 and y_2 are synchronized. In Fig. 8.4e, the G-functions of $g_i(i = 5, 6)$ for the synchronization of state variables x_3 and y_3 are presented. Because of $g_5 < 0$ with $g_6 > 0$, the two state variables x_3 and y_3 are also synchronized. The trajectories for the slave Chua system are presented through (y_1, y_2) and (y_1, y_3) planes in Fig. 8.4f, g. The synchronization domain of the slave system with $(y_3 = 0)$ is also superimposed in the (y_1, y_2) plane. In Fig. 8.4f, it is clearly observed that the synchronization vanishing exists on the boundaries between the synchronization and non-synchronization domains, and the synchronization appearance is in the synchronization domain. The trajectory of the slave Chua system in Fig. 8.4g is

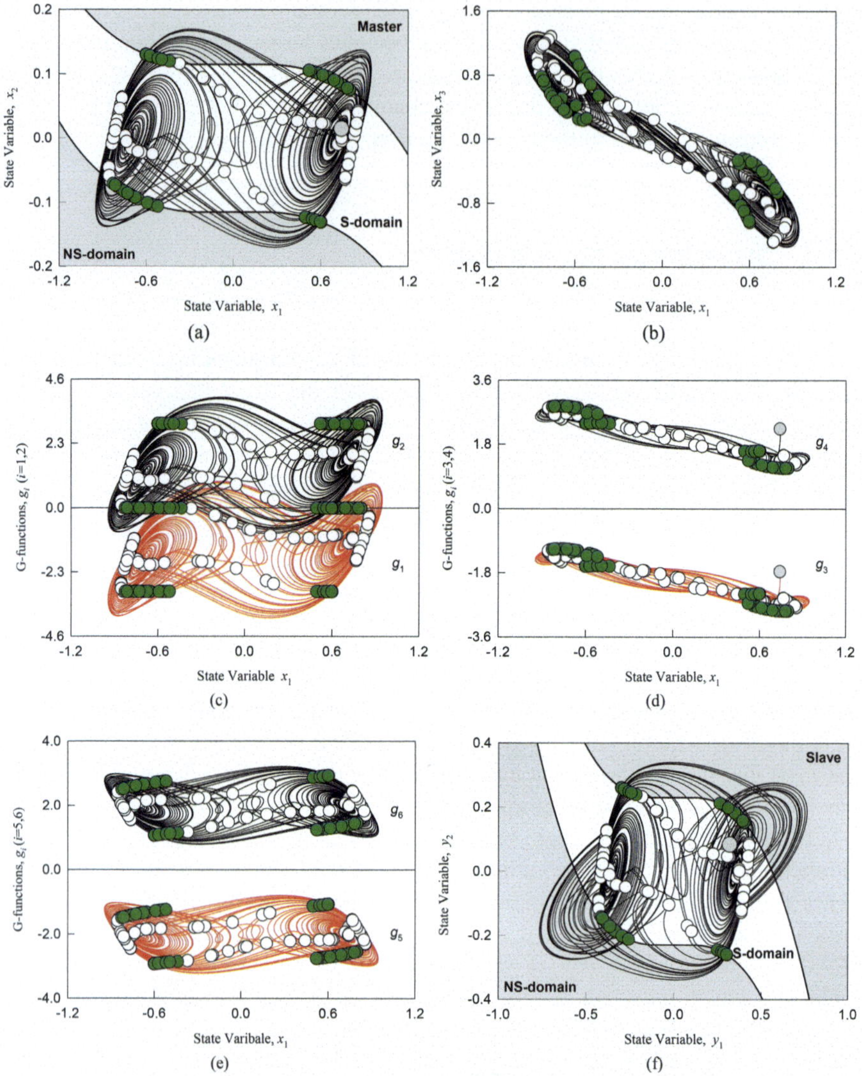

Fig. 8.4 The partial projective synchronicity for two Chua systems: (**a**) (x_1, x_2) with the invariant domain ($x_3 = 0$); (**b**) (x_1, x_3); (**c**) $(x_1, g_{1,\ 2})$; (**d**) $(x_1, g_{3,\ 4})$; (**e**) $(x_1, g_{5,\ 6})$; (**f**) (y_1, y_2) with the invariant domain ($y_3 = 0$); (**g**) (y_1, y_3); (**h**) (x_1, y_1); (**i**) (x_2, y_2); and (**j**) (x_3, y_3). (Hollow circular symbols: synchronization appearance; filled circular symbols: synchronization disappearance.) ($p_1 = 0.5$, $p_2 = 2.0$, $p_3 = 1.5$ and $k_1 = 1.5$, $k_2 = 2.0$, $k_3 = 2.0$)

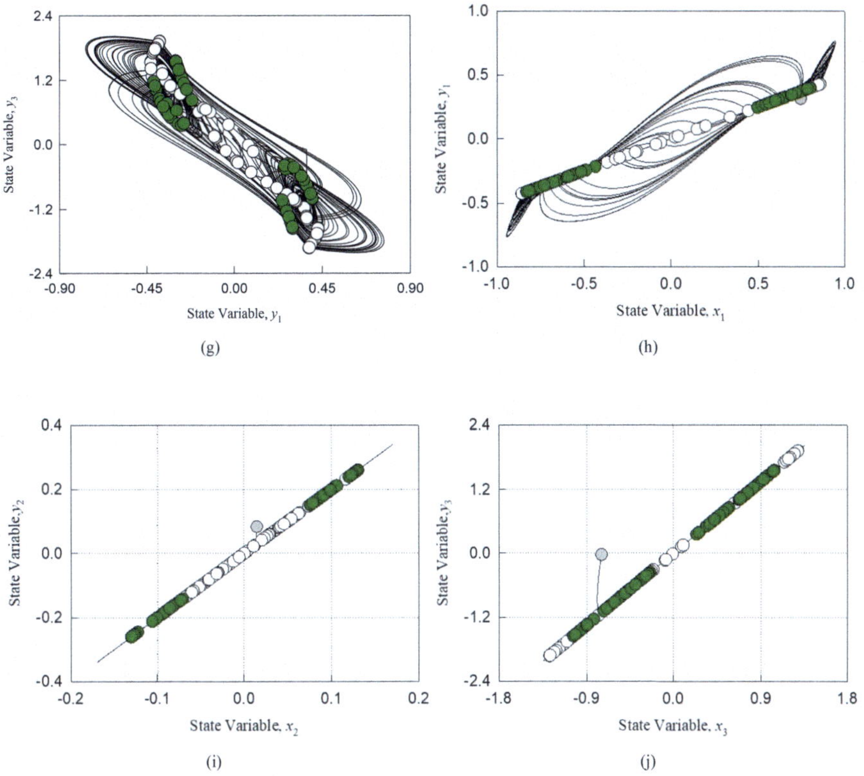

Fig. 8.4 (continued)

without similarity to the trajectory of the master Chua system in Fig. 8.4b with projective scales due to the non-synchronization. In Fig. 8.4h–j, the projective synchronizations are presented through the (x_1, y_1), (x_2, y_2), and (x_3, y_3) planes, respectively. The state variables (x_2, x_3) and (y_2, y_3) are synchronized with projective scales p_2 and p_3, but the state variables x_1 and y_1 are partially synchronized with the projective scale p_1.

For a clear observation of the projective synchronization of the master and slave Chua systems, the time histories of state variables and G-functions for the afore-discussed master and slave systems are presented in Fig. 8.4. The shaded and white areas are for the projective synchronization and non-synchronization, respectively. The time histories of the master and slave state variables x_i and y_i $(i = 1, 2, 3)$ are presented in Fig. 8.5a–c, respectively. In Fig. 8.5a, in the synchronization portions, the two state variables are synchronized with the projective scale p_1, but in the non-synchronization part, the two state variables x_1 and y_1 satisfy $x_1 = p_1 y_1$ for synchronization. In Fig. 8.5b, c, the state variables x_i and y_i $(i = 2, 3)$ possess the

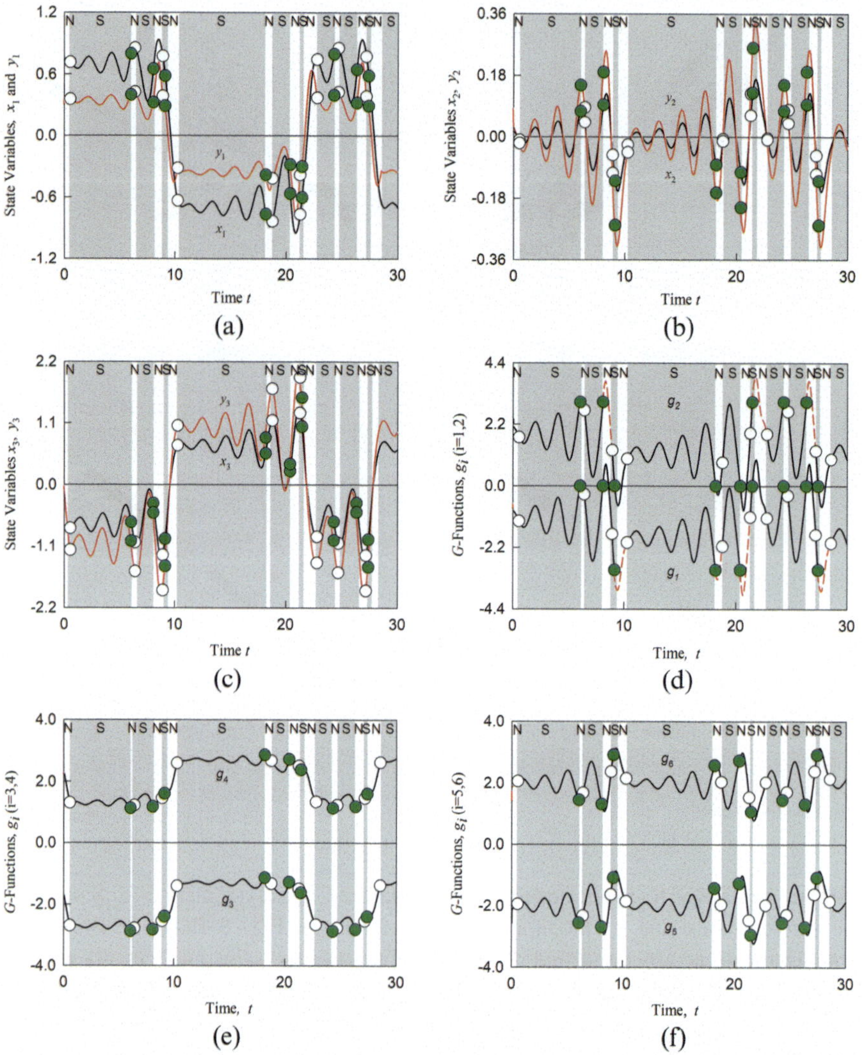

Fig. 8.5 The time histories for the partial projective synchronization of two Chua circuits. Responses: **(a)** (t, x_1) and (t, y_1), **(b)** (t_1, x_2) and (t, y_2), **(c)** (t, x_3) and (t, y_3). G-functions: **(d)** $(t, g_{1,2})$, **(e)**$(t, g_{3,4})$, and **(f)** $(t, g_{5,6})$. (Hollow circular symbols: synchronization appearance; filled circular symbols: synchronization disappearance.) ($p_1 = 0.5$, $p_2 = 2.0$, $p_3 = 1.5$ and $k_1 = 1.5$, $k_2 = 2.0$, $k_3 = 2.0$)

relation of $x_i = p_i y_i$ for the projective synchronization. Such statements can be verified through the G-functions in the analytical conditions of projective synchronization, as developed in the previous section. In Fig. 8.5d, f, presented are the time histories of G-functions, g_{2i-1} and g_{2i} $(i = 1, 2, 3)$, respectively. From the time histories of G-functions in Fig. 8.5d, the projective synchronization of the two state

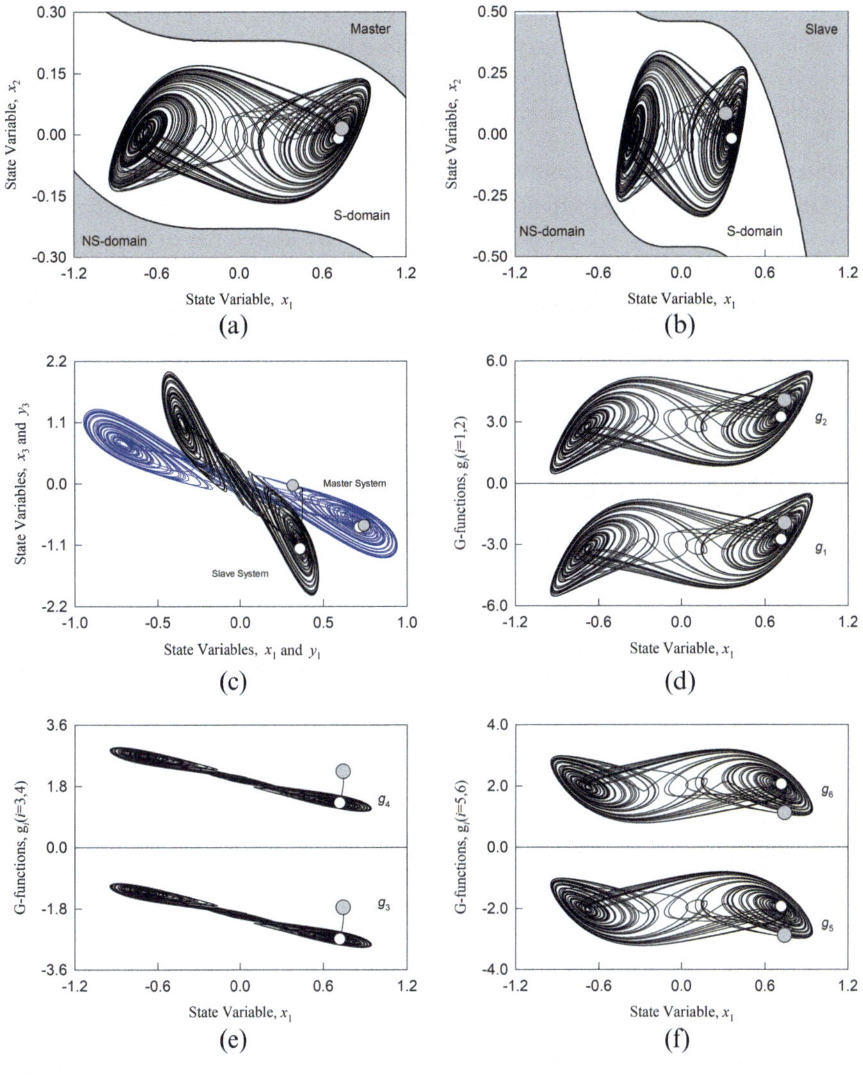

Fig. 8.6 The full projective synchronization for two Chua circuits with the different motions. Trajectories: (**a**)$x_1 - x_2$ plane with the invariant domain (master), (**b**) $y_1 - y_2$ plane with the invariant domain (slave), (**c**) $x_1 - x_3$ plane with $y_1 - y_3$ plane. G-functions: (**d**) $g_{1,\,2}$ versus x_1, (**e**) $g_{3,\,4}$ versus x_1, (**f**) $g_{5,\,6}$ versus x_1 (hollow circular symbols: synchronization appearance; gray symbols: the initial conditions). ($p_1 = 0.5$, $p_2 = 2.0$, $p_3 = 1.5$, and $k_1 = 3.0$, $k_2 = 2.0$, $k_3 = 2.0$)

variables x_1 and y_1 possesses the G-functions of $g_1 < 0$ and $g_2 > 0$, and the non-synchronization of the two state variables x_1 and y_1 has the G-functions of $g_1 < 0$ and $g_2 > 0$. From the time histories of G-functions in Fig. 8.5e, the G-functions of $g_3 < 0$ and $g_4 > 0$ hold for the projective synchronization of the two state variables x_2 and y_2. In Fig. 8.5f, the G-functions of $g_5 < 0$ and $g_6 > 0$

always exist for the synchronization of state variables x_3 and y_3. The switching points given by circular symbols satisfy the vanishing and appearance conditions of the projective synchronization. The green and hollow symbols are for synchronization vanishing and appearance, respectively.

To obtain the full projective synchronization for the controlled circuit with the double scroll chaotic circuit, the control parameters for $k_1 = 3.0$ and $k_2 = k_3 = 2.0$ are selected for numerical simulations, as shown in Fig. 8.6. The projective scale factors are $p_1 = 0.5$, $p_2 = 2.0$, and $p_3 = 1.5$. The phase trajectories of the master and slave systems are plotted through the (x_1, x_2) and (y_1, y_2) planes in Fig. 8.6a, b, and the corresponding invariant domains with $x_3 = 0$ and $y_3 = 0$ are also embedded in the two planes. All the trajectories of the master and slave Chua systems lie in the corresponding synchronization invariant domain. The trajectories in (x_1, x_3) and (y_1, y_3) planes for the master and slave systems are plotted together for the synchronization with the specific projective scale factors. The G-function distributions of (g_{2i-1}, g_{2i}) $(i = 1, 2, 3)$ versus the state variable x_1 are presented in Fig. 8.6d–f. $g_{2i-1} < 0$ and $g_{2i} > 0$ $(i = 1, 2, 3)$ are observed in the three plots. The analytical conditions of the projective synchronization are satisfied. Thus, the two Chua systems under control are fully synchronized. In order to compare with experimental results, the trajectories in $\{(x_1, x_2)$ and $(y_1, y_2)\}$, $\{(x_2, x_3)$ and $(y_2, y_3)\}$, and $\{(x_1, x_3)$ and $(y_1, y_3)\}$ for the master and slave systems are presented with specific projective scale factors in Fig. 8.7a–f, respectively.

8.6 Circuit Implementation

As in Min and Luo (2015), two coupled Chua circuits will be developed for their projective synchronization. The two circuits of master system (Eq. 8.1) and the controlled slave system (Eq. 8.8) are presented in Fig. 8.9a, b, respectively. The equation of circuits for the master system in Eq. 8.1 is

$$\begin{aligned} \frac{dx_1}{dt} &= \frac{1}{R_4C_1}\left[\frac{R_3}{R_1}x_2 - \frac{R_3}{R_2}h(x_1)\right], \\ \frac{dx_2}{dt} &= \frac{1}{R_{11}C_2}\left(\frac{R_{10}}{R_7}x_1 - \frac{R_{10}}{R_8}x_2 + \frac{R_{10}}{R_9}x_3\right), \\ \frac{dx_3}{dt} &= \frac{1}{R_{16}C_3}\left(-\frac{R_{15}}{R_{14}}x_2\right), \end{aligned} \tag{8.96}$$

with the nonlinearity $h(x_1)$ given by

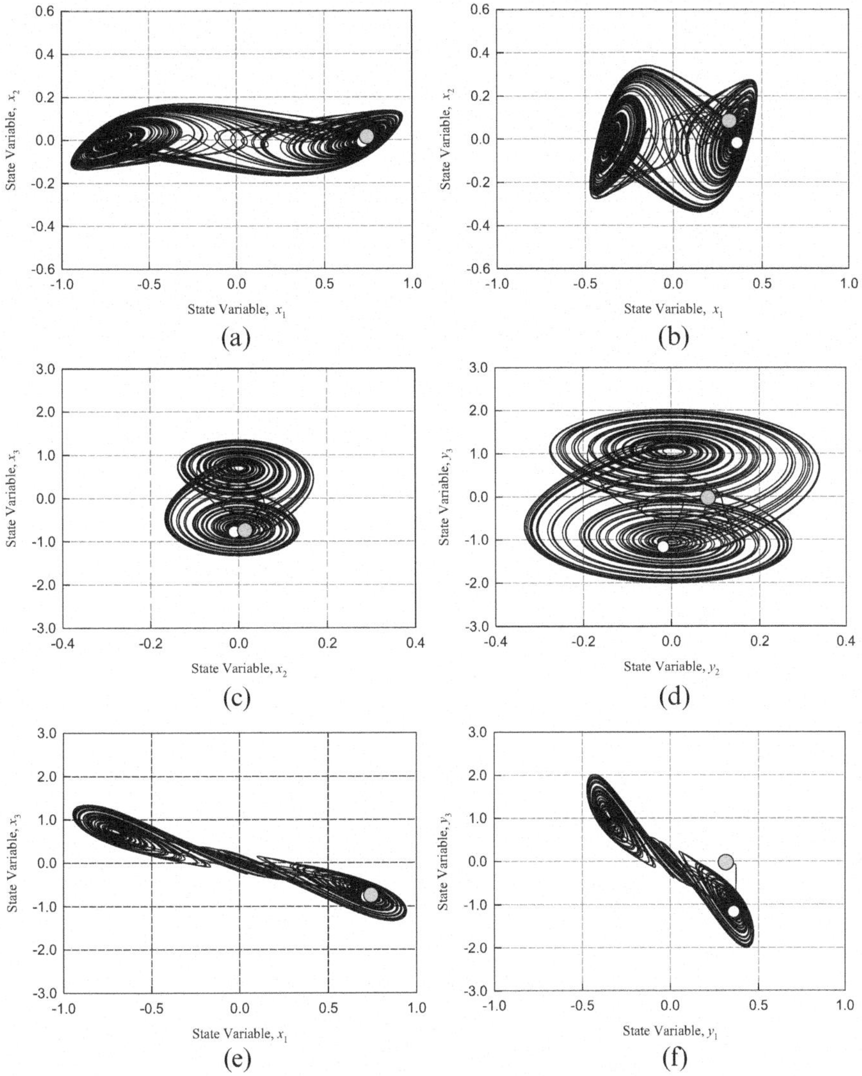

Fig. 8.7 The full projective synchronization for two Chua circuits. Trajectories: (**a**)(x_1, x_2) plane (master), (**b**) (y_1, y_2) plane (slave), (**c**) (x_2, x_3) plane (master), (**d**) (y_2, y_3) plane (slave), (**e**)(x_1, x_3) plane (master), (**f**) (y_1, y_3) plane (slave). (Hollow circular symbols: synchronization appearance; gray symbols: the initial conditions). ($p_1 = 0.5, p_2 = 2.0, p_3 = 1.5$, and $k_1 = 3.0, k_2 = 2.0, k_3 = 2.0$)

$$h(x_1) = -\frac{R_{19}}{R_{20}}x_1 + \frac{R_{19}}{R_{21}}x_1^3. \tag{8.97}$$

Comparison of Eqs. 8.1 and 8.96 gives

$$\alpha_1 = \frac{R_3}{R_1} = \frac{R_3}{R_2}, \beta_1 = \frac{R_{15}}{R_{14}}, a = -\frac{R_{19}}{R_{20}}, b = \frac{R_{19}}{R_{21}}. \tag{8.98}$$

The circuit parameters of Eq. 8.96 are $R_i = 10\ \text{k}\Omega$, $(i = 1, 2, 5, 6, 7, 8, 9, 10, 12, 13, 17, 18)$, $R_{19} = 2\ \text{k}\Omega$, $R_3 = R_{15} = 100\ \text{k}\Omega$, $R_4 = R_{11} = R_{16} = 50\text{k}\Omega$,$R_{14} = R_{21} = 7\text{k}\Omega$, ,$R_{20} = 14\ \text{k}\Omega$, $C_1 = C_2 = C_3 = 20\ \mu$ F to obtain the specific parameters of $\alpha_1\beta_1$, a and b.

The equation of circuits for the controlled slave system is

$$\begin{aligned}
\frac{dy_1}{dt} &= \frac{1}{R_{25}C_4}\left(\frac{R_{24}}{R_{23}}x_2 - \frac{R_{24}}{R_{22}}h(y_1) - \frac{R_{24}}{R_{51}}u_1\right), \\
\frac{dy_2}{dt} &= \frac{1}{R_{31}C_5}\left(\frac{R_{30}}{R_{34}}x_1 - \frac{R_{30}}{R_{29}}x_2 + \frac{R_{30}}{R_{28}}x_3 - \frac{R_{30}}{R_{60}}u_2\right), \\
\frac{dy_3}{dt} &= \frac{1}{R_{37}C_6}\left(-\frac{R_{36}}{R_{35}}x_2 - \frac{R_{36}}{R_{69}}u_3\right)
\end{aligned} \tag{8.99}$$

with

$$h(y_1) = -\frac{R_{42}}{R_{41}}x_1 + \frac{R_{42}}{R_{40}}x_1^3, \tag{8.100}$$

and the circuit of the control law in Eq. 8.5 is given by

$$\begin{aligned}
u_1 &= \frac{R_{24}}{R_{55}}\operatorname{sgn}\left[\frac{R_{50}}{R_{47}}y_1 - \left(1 + \frac{R_{50}}{R_{47}}\right)\frac{R_{44}R_{46}R_{49}}{(R_{48}+R_{49})R_{43}R_{45}}x_1\right], \\
u_2 &= \frac{R_{30}}{R_{68}}\operatorname{sgn}\left[\frac{R_{63}}{R_{60}}y_2 - \left(1 + \frac{R_{63}}{R_{60}}\right)\frac{R_{57}R_{59}R_{62}}{(R_{61}+R_{62})R_{56}R_{58}}x_2\right], \\
u_3 &= \frac{R_{36}}{R_{81}}\operatorname{sgn}\left[\frac{R_{76}}{R_{73}}y_3 - \left(1 + \frac{R_{76}}{R_{73}}\right)\frac{R_{70}R_{72}R_{75}}{(R_{74}+R_{75})R_{69}R_{71}}x_3\right].
\end{aligned} \tag{8.101}$$

Comparing Eq. 8.8 and Eqs. 8.99–8.101, the following equations are given by

$$\begin{aligned}
&\alpha_2 = \frac{R_{24}}{R_{23}} = \frac{R_{24}}{R_{22}}, \beta_2 = \frac{R_{36}}{R_{35}}, a = -\frac{R_{42}}{R_{41}}, b = \frac{R_{42}}{R_{40}}, \\
&k_1 = \frac{R_{24}}{R_{55}}, k_2 = \frac{R_{30}}{R_{68}}, k_3 = \frac{R_{36}}{R_{81}}, \\
&p_1 = \frac{R_{44}}{R_{43}}, p_2 = \frac{R_{57}}{R_{56}}, p_3 = \frac{R_{70}}{R_{69}}.
\end{aligned} \tag{8.102}$$

The circuit parameters in Eq. 8.99 are $R_i = 10\text{k}\Omega$, $(i = 22, 23, 26, 27, 28, 29, 30, 32, 33, 34, 38, 39)$, $R_{24} = 90$ kΩ, $R_{25} = R_{31} = R_{37} = 50$ kΩ,$R_{35} = R_{40} = 7$ kΩ,$R_{36} = 100$ kΩ, $R_{42} = 2$ kΩ,$R_{41} = 14$ kΩ, $C_4 = C_5 = C_6 = 20$ μF. And the circuit parameters of the control law in Eq. 8.101 are $R_i = 10$ kΩ, $(i = 43, 45, \cdots, 52, 54, 56, 58, \cdots, 65, 67, 69, 71, \cdots, 78, 80)$, $R_{53} = R_{66} = R_{79} = 125$ kΩ.

The designed circuits of the master and controlled slave systems in Fig. 8.8 have three channels to conduct the integration of three state variables, respectively. The operational amplifiers TL082CP, resistors, and capacitors perform the basic operations of addition, subtraction, and integration. The nonlinear terms in Eq. 8.2 or Eq. 8.4 are completed by two multipliers AD633JN, an amplifier, and three resistors. Here, the voltage supplies of all active devices are of ± 15 V, as shown in Fig. 8.8. The circuit of the sign function in Eq. 8.101 is realized by the null detector, which is a type of amplifier for null comparison measurements.

The adjustments of three scaling factors p_i $(i = 1, 2, 3)$ are performed by adjusting R_{44}, R_{57}, and R_{70}, respectively. From Eq. 8.95, the scaling factors of $p_1 = 0.5$, $p_2 = 2.0$, and $p_3 = 1.5$ correspond to $R_{44} = 5$ kΩ, $R_{57} = 20$ kΩ, and $R_{70} = 15$ kΩ, respectively. The control parameters k_i $(i = 1, 2, 3)$ vary with the resistors of R_{55}, R_{68}, and R_{81}, respectively. The full projective synchronization for the controlled slave Chua circuit with the master Chua circuits with the double scroll chaotic behaviors is carried out experimentally. For $k_1 = 3.0$ and $k_2 = k_3 = 2.0$, the resistors of $R_{55} = 30$ kΩ, $R_{68} = 5$ kΩ, and $R_{81} = 50$ kΩ are used in circuits for a full projective synchronization. The trajectories for the full projective synchronization of the two Chua circuits with different behaviors are generated by the circuits, as shown in Fig. 8.9. The scales of horizontal and vertical axes are 0.5 V/div and 0.2 V/div in (x_1, x_2) and (y_1, y_2) planes, 0.2 V/div and 1 V/div in (x_2, x_3) and (y_2, y_3) planes, and 0.5 V/div and 1 V/div in (x_1, x_3) and (y_1, y_3) planes, respectively. From the trajectories in Figs. 8.7 and 8.9, numerical and experimental results are in a good agreement. The full projective synchronization of the two Chua systems can be realized through the Chua circuits. Similarly, partial projective synchronization of the two Chua systems can be done experimentally.

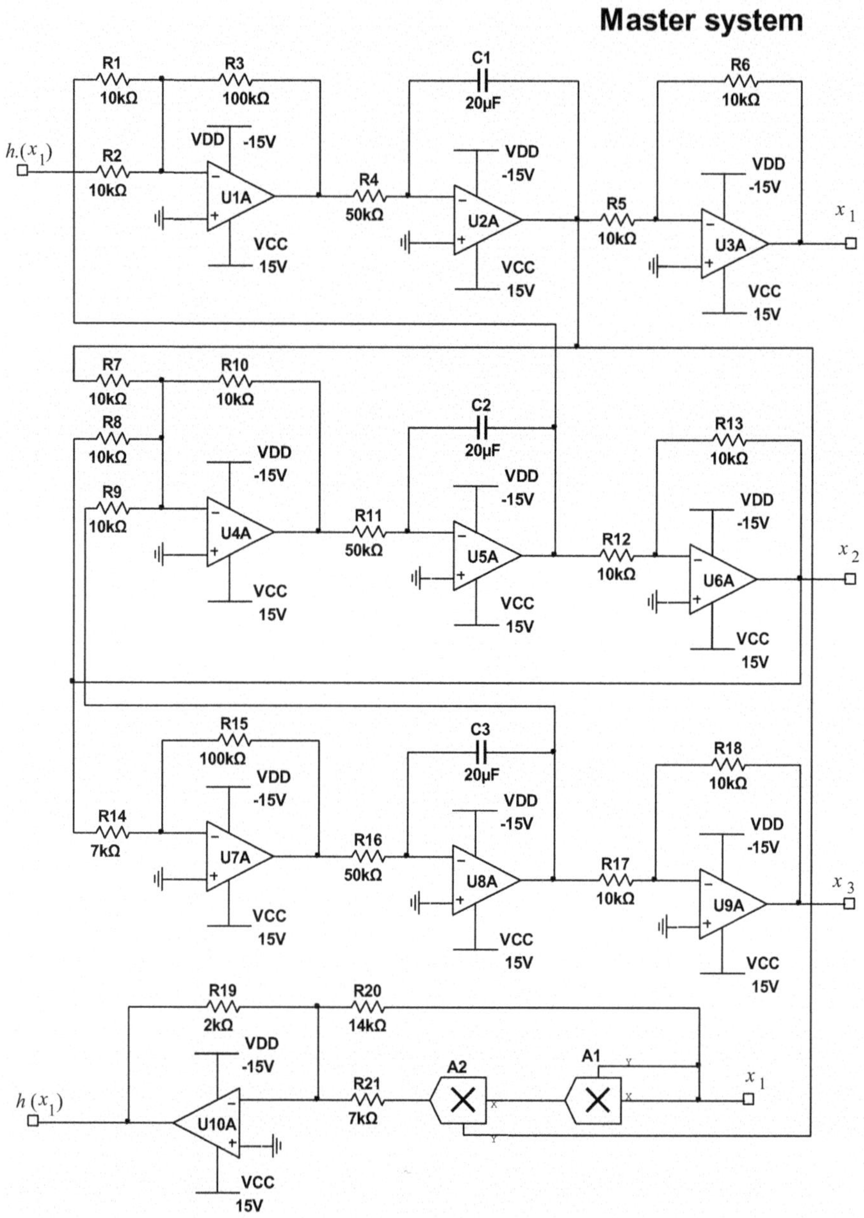

(a)

Fig. 8.8 Circuit sketches for the projective synchronization between the master and controlled slave systems: (**a**) master system, (**b**) the controller and slave system

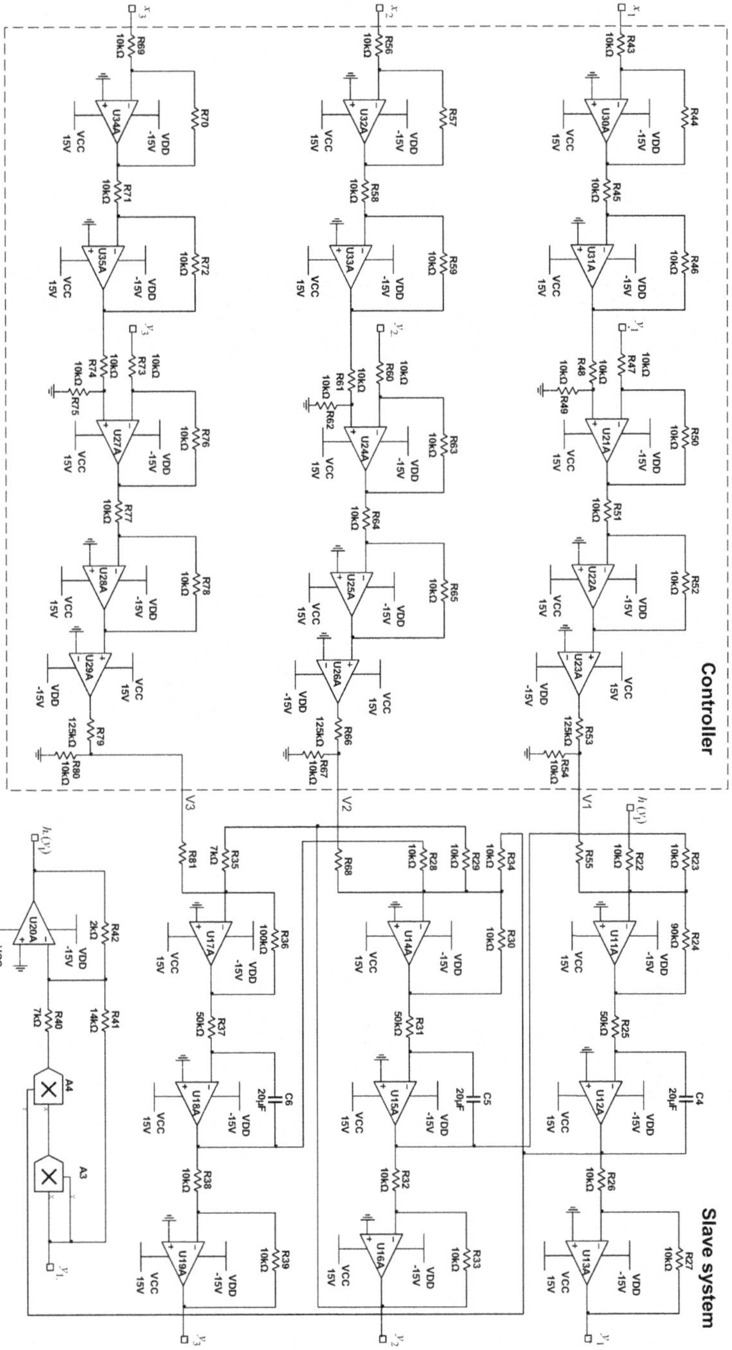

(b)

Fig. 8.8 (continued)

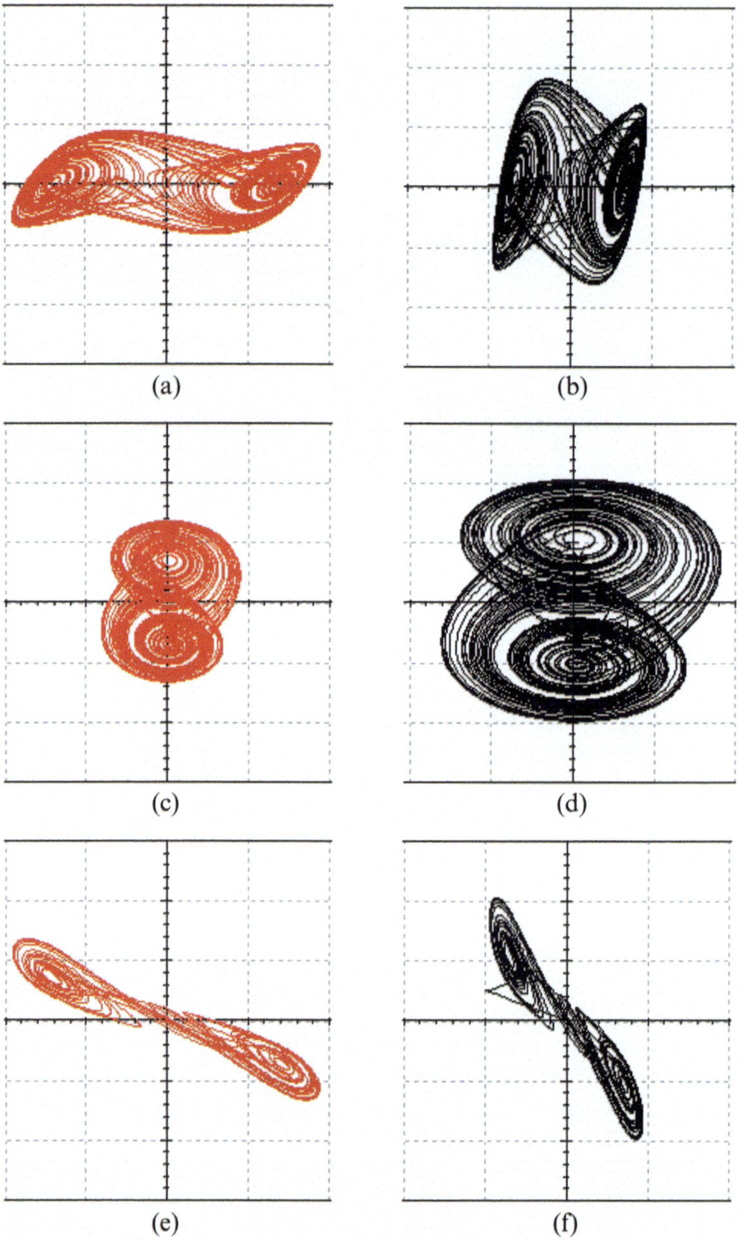

Fig. 8.9 The experimental output trajectories of full synchronization between the master and slave Chua circuits. (**a**, **b**) (x_1, x_2) and (y_1, y_2) with 0.5v/div (horizontal) and 0.2 v/div (vertical); (**c**, **d**) (x_2, x_3) and (y_2, y_3) with 0.2 v/div (horizontal) and 1 v/div (vertical); (**e**, **f**) (x_1, x_3) and (y_1, y_3) with 0.5 v/div (horizontal) and 1 v/div (vertical). ($R_{44} = 5\ \text{k}\Omega$, $R_{55} = 30\ \text{k}\Omega$, $R_{57} = 20\ \text{k}\Omega$, $R_{68} = 5\ \text{k}\Omega$, $R_{70} = 15\ \text{k}\Omega$, $R_{81} = 50\ \text{k}\Omega$)

References

Luo, A. C. J. (2009a). A theory for synchronization of dynamical systems. *Communications in Nonlinear Science and Numerical Simulation, 14*, 1901–1951.
Luo, A. C. J. (2009b). *Discontinuous dynamical systems on time-varying domains*. Higher Education Press/Springer.
Luo, A. C. J. (2012). *Discontinuous dynamical systems*. Higher Education Press/Springer.
Luo, A. C. J. (2013). *Dynamical system synchronization*. Springer.
Min, F. H., & Luo, A. C. J. (2015). Complex dynamics of projective synchronization of Chua circuits with different scrolls. *International Journal of Bifurcation and Chaos, 25*(5), 1530016.

Chapter 9
Function Synchronizations of Dynamic Systems

Contents

9.1 Description of Systems 151
9.2 Switching Synchronization Dynamics 155
9.3 Switching Projective 159
9.4 Synchronization Scenario and Parameter Maps 163
9.5 Numerical Illustration 164
References 170

In this chapter, the function synchronization of the controlled Duffing oscillator with van der Pol system and pendulum system is studied. Based on the theory of discontinuous dynamical systems, the analytical conditions for chaos synchronization of three different systems are obtained under sinusoidal constraints. With these conditions, the control parameter maps and the synchronization invariant sets are derived, respectively. Numerical illustrations for partial and full combination synchronization among three systems are carried out.

9.1 Description of Systems

As in Min and Ma (2020), consider three systems as follows:

The van der Pol is expressed by

$$\begin{aligned}&\dot{\mathbf{x}} = \mathbf{F}(\mathbf{x}, t), \text{with}\\&x_2 \equiv \dot{x}_1, F(\mathbf{x}, t) = x_2\left(1 - x_1^2\right)b - x_1 + Q\sin(\Omega t).\end{aligned} \tag{9.1}$$

The periodically driven pendulum

F. Min, *Discontinuous Dynamics and System Synchronization*, Nonlinear Systems and Complexity 39, https://doi.org/10.1007/978-3-031-66648-3_9

$$\begin{aligned}&\dot{\mathbf{y}} = \mathbf{F}(\mathbf{y}, t), \text{ with}\\ &y_2 \equiv \dot{y}_1, F(\mathbf{y}, t) = -\alpha_0 \sin y_1 + Q_0 \cos \Omega_0 t.\end{aligned} \tag{9.2}$$

and the Duffing system

$$\begin{aligned}&\dot{\mathbf{w}} = \mathbf{F}(\mathbf{w}, t), \text{ with}\\ &w_2 \equiv \dot{w}_1, F(\mathbf{w}, t) = -\delta w_2 + \alpha_1 w_1 - \alpha_2 w_1^3 + Q_1 \cos \Omega_1 t.\end{aligned} \tag{9.3}$$

where x_1, y_1, and w_1 represent the displacement of three systems and x_2, y_2, and w_2 indicate the velocity of van der Pol, pendulum, and Duffing, respectively.

For the switching combination synchronization of the slave system with two master systems, the controlled constraints are added to the slave system, which makes the controlled Duffing oscillator become discontinuous. Equation 9.3 is modified as

$$\dot{\mathbf{w}} = \mathbf{F}(\mathbf{w}, t) - \mathbf{u}(\mathbf{x}, \mathbf{w}, t), \tag{9.4}$$

where nonlinear feedback control laws are shown

$$\begin{aligned}&\mathbf{u}(\mathbf{x}, \mathbf{w}, t) = (u_1, u_2)^{\mathrm{T}}\\ &\text{with} \begin{cases} u_1 = k_1 \operatorname{sgn}(w_1 - \sin(p_1 x_1 + p_2 y_1)),\\ u_2 = k_2 \operatorname{sgn}(w_2 - (p_1 x_2 + p_2 y_2)\cos(p_1 x_1 + p_2 y_1)).\end{cases}\end{aligned} \tag{9.5}$$

where k_1, k_2 are control parameters and p_1, p_2 are scaling factors.

With the feedback switching controllers in Eq. 9.5, the controlled Duffing oscillator is divided into four motion domains and four boundaries. The corresponding vector fields are defined as $\mathbf{F}(\mathbf{w}, t) = (\mathrm{F}_1(\mathbf{w}, t), \mathrm{F}_2(\mathbf{w}, t))^{\mathrm{T}}$, described as follows:

(1) If $w_1 > \sin(p_1 x_1 + p_2 y_1)$ and $w_2 > (p_1 x_2 + p_2 y_2) \cos(p_1 x_1 + p_2 y_1)$,

$$\begin{aligned}\mathrm{F}_1(\mathbf{w}, t) &= w_2 - (p_1 x_2 + p_2 y_2) \cos(p_1 x_1 + p_2 y_1) - k_1\\ &\quad (-\delta w_2 + \alpha_1 w_1 - \alpha_2 w_1^3 + Q_1 \cos \Omega_1 t - k_2)\\ \mathrm{F}_2(\mathbf{w}, t) &= +(p_1 x_2 + p_2 y_2)^2 \sin(p_1 x_1 + p_2 y_1)\\ &\quad - \left[\left(x_2(1 - x_1^2)b - x_1 + Q \sin(\Omega t)\right) + (-\alpha_0 \sin y_1 + Q_0 \cos \Omega_0 t)\right] \cos(p_1 x_1 + p_2 y_1);\end{aligned} \tag{9.6}$$

(2) If $w_1 > \sin(p_1 x_1 + p_2 y_1)$and $w_2 > (p_1 x_2 + p_2 y_2) \cos(p_1 x_1 + p_2 y_1)$, $\mathrm{x}_2 > \mathrm{y}_2$

$$
\begin{aligned}
F_1(\mathbf{w},t) = {} & w_2 - (p_1x_2 + p_2y_2)\cos(p_1x_1 + p_2y_1) - k_1 \\
& \left(-\delta w_2 + \alpha_1 w_1 - \alpha_2 w_1^3 + Q_1 \cos\Omega_1 t + k_2\right) \\
F_2(\mathbf{w},t) = {} & +(p_1x_2 + p_2y_2)^2 \sin(p_1x_1 + p_2y_1) \\
& - \left[\left(x_2\left(1 - x_1^2\right)b - x_1 + Q\sin(\Omega t)\right) + (-\alpha_0 \sin y_1 + Q_0 \cos\Omega_0 t)\right]\cos(p_1x_1 + p_2y_1);
\end{aligned}
\tag{9.7}
$$

(3) If $w_1 >$ sin $(p_1x_1 + p_2y_1)$ and $w_2 > (p_1x_2 + p_2y_2)$ cos $(p_1x_1 + p_2y_1)$, $x_2 > y_2$

$$
\begin{aligned}
F_1(\mathbf{w},t) = {} & w_2 - (p_1x_2 + p_2y_2)\cos(p_1x_1 + p_2y_1) + k_1 \\
& \left(-\delta w_2 + \alpha_1 w_1 - \alpha_2 w_1^3 + Q_1 \cos\Omega_1 t + k_2\right) \\
F_2(\mathbf{w},t) = {} & +(p_1x_2 + p_2y_2)^2 \sin(p_1x_1 + p_2y_1) \\
& - \left[\left(x_2\left(1 - x_1^2\right)b - x_1 + Q\sin(\Omega t)\right) + (-\alpha_0 \sin y_1 + Q_0 \cos\Omega_0 t)\right]\cos(p_1x_1 + p_2y_1);
\end{aligned}
\tag{9.8}
$$

(4) If $w_1 >$ sin $(p_1x_1 + p_2y_1)$ and $w_2 > (p_1x_2 + p_2y_2)$ cos $(p_1x_1 + p_2y_1)$,

$$
\begin{aligned}
F_1(\mathbf{w},t) = {} & w_2 - (p_1x_2 + p_2y_2)\cos(p_1x_1 + p_2y_1) + k_1 \\
& \left(-\delta w_2 + \alpha_1 w_1 - \alpha_2 w_1^3 + Q_1 \cos\Omega_1 t - k_2\right) \\
F_2(\mathbf{w},t) = {} & +(p_1x_2 + p_2y_2)^2 \sin(p_1x_1 + p_2y_1) \\
& - \left[\left(x_2\left(1 - x_1^2\right)b - x_1 + Q\sin(\Omega t)\right) + (-\alpha_0 \sin y_1 + Q_0 \cos\Omega_0 t)\right]\cos(p_1x_1 + p_2y_1);
\end{aligned}
\tag{9.9}
$$

Constrained by Eq. 9.5, the boundaries varying with time will make the synchronization analysis difficult. Therefore, the relative coordinates are used:

$$
\begin{aligned}
z_1 &= w_1 - \sin(p_1x_1 + p_2y_1), \\
z_2 &= w_2 - (p_1x_2 + p_2y_2)\cos(p_1x_1 + p_2y_1).
\end{aligned}
\tag{9.10}
$$

The corresponding regions and boundaries become

$$
\begin{aligned}
&\Omega_1(t) = \{(z_1, z_2) | z_1 > 0, z_2 > 0\}, \\
&\Omega_2(t) = \{(z_1, z_2) | z_1 > 0, z_2 < 0\}, \\
&\Omega_3(t) = \{(z_1, z_2) | z_1 < 0, z_2 < 0\}, \\
&\Omega_4(t) = \{(z_1, z_2) | z_1 < 0, z_2 > 0\}; \\
&\partial\Omega_{12}(t) = \{(z_1, z_2) | z_2 = 0, z_1 > 0\}, \\
&\partial\Omega_{23}(t) = \{(z_1, z_2) | z_1 = 0, z_2 < 0\}, \\
&\partial\Omega_{34}(t) = \{(z_1, z_2) | z_2 = 0, z_1 < 0\}, \\
&\partial\Omega_{14}(t) = \{(z_1, z_2) | z_1 = 0, z_2 > 0\}.
\end{aligned}
\tag{9.11}
$$

By using Eq. 9.10, the boundaries in the relative coordinates become constant. Therefore, the domains and boundaries for the controlled Duffing oscillator can be depicted as shown in Fig. 9.1, where the dashed straight lines are the boundaries,

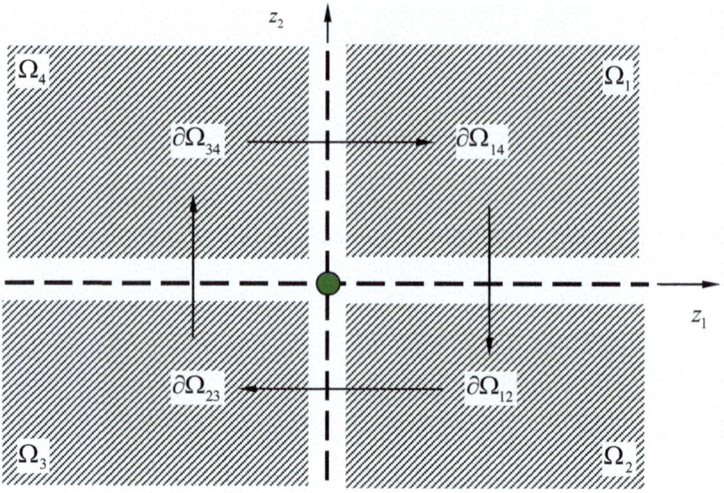

Fig. 9.1 Phase plane partitions and boundaries in relative coordinates

which are independent of time. Subsequently, with the relative coordinates, the analytical conditions for the coexistence of chaotic synchronization for three systems can be easily carried out.

The controlled slave system in the relative coordinates can be described as

$$\dot{\mathbf{z}}^{(\alpha)} = \mathbf{g}^{(\alpha)}\left(\mathbf{z}^{(\alpha)}, \mathbf{x}(t), t\right) \text{ with } \dot{\mathbf{x}} = \mathbf{F}(\mathbf{x}, t) \tag{9.12}$$

where

$$\begin{aligned}
g_1{}^{(\alpha)}\left(\mathbf{z}^{(\alpha)}, \mathbf{x}, t\right) &= z_2{}^{(\alpha)} - k_1 \text{ for } \alpha = 1,2;\\
g_1{}^{(\alpha)}\left(\mathbf{z}^{(\alpha)}, \mathbf{x}, t\right) &= z_2{}^{(\alpha)} + k_1 \text{ for } \alpha = 3,4;\\
g_2{}^{(\alpha)}\left(\mathbf{z}^{(\alpha)}, \mathbf{x}, t\right) &= G_w\left(\mathbf{z}^{(\alpha)}, \mathbf{x}, t\right) - k_2 + (p_1 x_2 + p_2 y_2)^2 \sin(p_1 x_1 + p_2 y_1)\\
&\quad - \left[p_1\left(x_2\left(1 - x_1^2\right)b - x_1 + Q\sin(\Omega t)\right) + p_2\left(-\alpha_0 \sin y_1\right.\right.\\
&\quad \left.\left. + Q_0 \cos \Omega_0 t\right)\right] \cos(p_1 x_1 + p_2 y_1), \text{ for } \alpha = 1,4;\\
g_2{}^{(\alpha)}\left(\mathbf{z}^{(\alpha)}, \mathbf{x}, t\right) &= G_w\left(\mathbf{z}^{(\alpha)}, \mathbf{x}, t\right) + k_2 + (p_1 x_2 + p_2 y_2)^2 \sin(p_1 x_1 + p_2 y_1)\\
&\quad - \left[p_1\left(x_2\left(1 - x_1^2\right)b - x_1 + Q\sin(\Omega t)\right) + p_2\left(-\alpha_0 \sin y_1\right.\right.\\
&\quad \left.\left. + Q_0 \cos \Omega_0 t\right)\right] \cos(p_1 x_1 + p_2 y_1), \text{ for } \alpha = 2,3.
\end{aligned}$$

$$\mathbf{g}^{(\alpha)}\left(\mathbf{z}^{(\alpha)}, \mathbf{x}, t\right) = \left(g_1{}^{(\alpha)}, g_2{}^{(\alpha)}, g_3{}^{(\alpha)}, g_4{}^{(\alpha)}\right)^T \tag{9.13}$$

with

$$G_w\left(\mathbf{z}^{(\alpha)},\mathbf{x},t\right)=\big(-\delta(z_2+(p_1x_2+p_2y_2)\cos(p_1x_1+p_2y_1))+\\ \alpha_1(z_1+\sin(p_1x_1+p_2y_1))-\alpha_2(z_1+\sin(p_1x_1+p_2y_1))^3\\ +Q_1\cos\Omega_1 t\big). \tag{9.14}$$

The governing equation on the boundary can be expressed by

$$\dot{\mathbf{z}}^{(\alpha\beta)}=\mathbf{g}^{(\alpha\beta)}\left(\mathbf{z}^{(\alpha\beta)},\mathbf{x}(t),t\right),\text{ with }\dot{\mathbf{x}}=\mathbf{F}(\mathbf{x},t) \tag{9.15}$$

where

$$g_1{}^{(\alpha\beta)}\left(\mathbf{z}^{(\alpha\beta)},\mathbf{x},t\right)=z_2=0\text{ and }g_2{}^{(\alpha\beta)}\left(z^{(\alpha\beta)},t\right)=0.\\ z_1{}^{(\alpha\beta)}=0\text{ and }z_2{}^{(\alpha\beta)}=0\text{ on }\partial\Omega_{\alpha\beta}\text{ for }(\alpha,\beta)=(2,3),(1,4),\\ z_1{}^{(\alpha\beta)}=c\text{ and }z_2{}^{(\alpha\beta)}=0\text{ on }\partial\Omega_{\alpha\beta}\text{ for }(\alpha,\beta)=(1,2),(3,4). \tag{9.16}$$

9.2 Switching Synchronization Dynamics

From the theory of discontinuous system in (Luo, 2008, 2009a, b, 2012), all the G-functions and corresponding high-order G-functions can be shown as

$$G^{(\alpha)}_{\partial\Omega_{ij}}(\mathbf{z}_m,\mathbf{x},t_{m\pm})=\mathbf{n}^T_{\partial\Omega_{ij}}\cdot\left[\mathbf{g}^{(\alpha)}(\mathbf{z}_m,\mathbf{x},t_{m\pm})-\mathbf{g}^{(ij)}(\mathbf{z}_m,\mathbf{x},t_{m\pm})\right],\\ G^{(1,\alpha)}_{\partial\Omega_{ij}}(\mathbf{z}_m,\mathbf{x},t_{m\pm})=\mathbf{n}^T_{\partial\Omega_{ij}}\cdot\left[D\mathbf{g}^{(\alpha)}(\mathbf{z}_m,\mathbf{x},t_{m\pm})-D\mathbf{g}^{(ij)}(\mathbf{z}_m,\mathbf{x},t_{m\pm})\right]. \tag{9.17}$$

where $G^{(\alpha)}_{\partial\Omega_{ij}}(\mathbf{z}_m,\mathbf{x},t_{m\pm})$ and $G^{(1,\alpha)}_{\partial\Omega_{ij}}(\mathbf{z}_m,\mathbf{x},t_{m\pm})$ are the zero-order and first-order G-functions of the flow in the domain $\Omega_\alpha(\alpha\in\{i,j\})$ at the boundary $\partial\Omega_{ij}(i,j=1,$ 2, 3, 4).

From Eq. 9.16, the normal vectors of the relative boundaries are

$$\mathbf{n}_{\partial\Omega_{12}}=\mathbf{n}_{\partial\Omega_{34}}=(0,1)^{\mathrm{T}}\text{ and }\mathbf{n}_{\partial\Omega_{14}}=\mathbf{n}_{\partial\Omega_{23}}=(1,0)^{\mathrm{T}}. \tag{9.18}$$

The G-functions at the boundary for eight domains $\Omega_\alpha(\alpha\in\{1,2,3,4\})$ are described as

$$
\begin{aligned}
&G_{\partial\Omega_{12}}^{(\alpha)}(\mathbf{z}_m, \mathbf{x}, t_{m\pm}) = G_{\partial\Omega_{34}}^{(\alpha)}(\mathbf{z}_m, \mathbf{x}, t_{m\pm}) = g_2{}^{(\alpha)}(\mathbf{z}_m, \mathbf{x}, t_{m\pm}),\\
&G_{\partial\Omega_{23}}^{(\alpha)}(\mathbf{z}_m, \mathbf{x}, t_{m\pm}) = G_{\partial\Omega_{14}}^{(\alpha)}(\mathbf{z}_m, \mathbf{x}, t_{m\pm}) = g_1{}^{(\alpha)}(\mathbf{z}_m, \mathbf{x}, t_{m\pm});\\
&G_{\partial\Omega_{12}}^{(1,\alpha)}(\mathbf{z}_m, \mathbf{x}, t_{m\pm}) = G_{\partial\Omega_{34}}^{(1,\alpha)}(\mathbf{z}_m, \mathbf{x}, t_{m\pm}) = Dg_2{}^{(\alpha)}(\mathbf{z}_m, \mathbf{x}, t_{m\pm}),\\
&G_{\partial\Omega_{23}}^{(1,\alpha)}(\mathbf{z}_m, \mathbf{x}, t_{m\pm}) = G_{\partial\Omega_{14}}^{(1,\alpha)}(\mathbf{z}_m, \mathbf{x}, t_{m\pm}) = Dg_1{}^{(\alpha)}(\mathbf{z}_m, \mathbf{x}, t_{m\pm});
\end{aligned}
\tag{9.19}
$$

The G-functions in corresponding regions are presented by

$$
\begin{aligned}
&G_{\partial\Omega_{12}}^{(\alpha)}\left(\mathbf{z}^{(\alpha)}, \mathbf{x}, t\right) = G_{\partial\Omega_{34}}^{(\alpha)}\left(\mathbf{z}^{(\alpha)}, \mathbf{x}, t\right) = g_2\left(\mathbf{z}^{(\alpha)}, \mathbf{x}, t\right),\\
&G_{\partial\Omega_{23}}^{(\alpha)}\left(\mathbf{z}^{(\alpha)}, \mathbf{x}, t\right) = G_{\partial\Omega_{14}}^{(\alpha)}\left(\mathbf{z}^{(\alpha)}, \mathbf{x}, t\right) = g_1\left(\mathbf{z}^{(\alpha)}, \mathbf{x}, t\right).
\end{aligned}
\tag{9.20}
$$

From Luo (2008, 2009a, b, 2012), the criterions of a flow sliding on the relative boundaries of $\partial\Omega_{12}$, $\partial\Omega_{34}$, $\partial\Omega_{23}$ and $\partial\Omega_{14}$ for the controlled slave system are given by

$$
\begin{aligned}
&\left.\begin{aligned}
&G_{\partial\Omega_{12}}^{(1)}(\mathbf{z}_m, \mathbf{x}, t_{m-}) = g_2^{(1)}(\mathbf{z}_m, \mathbf{x}, t_{m-}) < 0,\\
&G_{\partial\Omega_{12}}^{(2)}(\mathbf{z}_m, \mathbf{x}, t_{m-}) = g_2^{(2)}(\mathbf{z}_m, \mathbf{x}, t_{m-}) > 0,
\end{aligned}\right\}\text{for } \mathbf{z}_m \in \partial\Omega_{12};\\
&\left.\begin{aligned}
&G_{\partial\Omega_{34}}^{(3)}(\mathbf{z}_m, \mathbf{x}, t_{m-}) = g_2^{(3)}(\mathbf{z}_m, \mathbf{x}, t_{m-}) > 0,\\
&G_{\partial\Omega_{34}}^{(4)}(\mathbf{z}_m, \mathbf{x}, t_{m-}) = g_2^{(4)}(\mathbf{z}_m, \mathbf{x}, t_{m-}) < 0,
\end{aligned}\right\}\text{for } \mathbf{z}_m \in \partial\Omega_{34};\\
&\left.\begin{aligned}
&G_{\partial\Omega_{23}}^{(2)}(\mathbf{z}_m, \mathbf{x}, t_{m-}) = g_1^{(2)}(\mathbf{z}_m, \mathbf{x}, t_{m-}) < 0,\\
&G_{\partial\Omega_{23}}^{(3)}(\mathbf{z}_m, \mathbf{x}, t_{m-}) = g_1^{(3)}(\mathbf{z}_m, \mathbf{x}, t_{m-}) > 0,
\end{aligned}\right\}\text{for } \mathbf{z}_m \in \partial\Omega_{23};\\
&\left.\begin{aligned}
&G_{\partial\Omega_{14}}^{(1)}(\mathbf{z}_m, \mathbf{x}, t_{m-}) = g_1^{(1)}(\mathbf{z}_m, \mathbf{x}, t_{m-}) < 0,\\
&G_{\partial\Omega_{14}}^{(4)}(\mathbf{z}_m, \mathbf{x}, t_{m-}) = g_1^{(4)}(\mathbf{z}_m, \mathbf{x}, t_{m-}) > 0,
\end{aligned}\right\}\text{for } \mathbf{z}_m \in \partial\Omega_{14}.
\end{aligned}
\tag{9.21}
$$

From Luo (2008, 2009a, b, 2012), the criterions of a passing flow through the relative boundaries of $\partial\Omega_{12}$, $\partial\Omega_{34}$, $\partial\Omega_{23}$ and $\partial\Omega_{14}$ for the controlled slave system are given by

$$\left.\begin{aligned}&G_{\partial\Omega_{12}}^{(1)}(\mathbf{z}_m,\mathbf{x},t_{m-})=g_2^{(1)}(\mathbf{z}_m,\mathbf{x},t_{m-})<0,\\&G_{\partial\Omega_{12}}^{(2)}(\mathbf{z}_m,\mathbf{x},t_{m-})=g_2^{(2)}(\mathbf{z}_m,\mathbf{x},t_{m+})<0,\end{aligned}\right\}\text{for } \mathbf{z}_m\in\partial\Omega_{12};$$
$$\left.\begin{aligned}&G_{\partial\Omega_{34}}^{(3)}(\mathbf{z}_m,\mathbf{x},t_{m-})=g_2^{(3)}(\mathbf{z}_m,\mathbf{x},t_{m-})>0,\\&G_{\partial\Omega_{34}}^{(4)}(\mathbf{z}_m,\mathbf{x},t_{m-})=g_2^{(4)}(\mathbf{z}_m,\mathbf{x},t_{m+})>0,\end{aligned}\right\}\text{for } \mathbf{z}_m\in\partial\Omega_{34};$$
$$\left.\begin{aligned}&G_{\partial\Omega_{23}}^{(2)}(\mathbf{z}_m,\mathbf{x},t_{m-})=g_1^{(2)}(\mathbf{z}_m,\mathbf{x},t_{m-})<0,\\&G_{\partial\Omega_{23}}^{(3)}(\mathbf{z}_m,\mathbf{x},t_{m-})=g_1^{(3)}(\mathbf{z}_m,\mathbf{x},t_{m+})<0,\end{aligned}\right\}\text{for } \mathbf{z}_m\in\partial\Omega_{23};$$
$$\left.\begin{aligned}&G_{\partial\Omega_{14}}^{(1)}(\mathbf{z}_m,\mathbf{x},t_{m-})=g_1^{(1)}(\mathbf{z}_m,\mathbf{x},t_{m-})>0,\\&G_{\partial\Omega_{14}}^{(4)}(\mathbf{z}_m,\mathbf{x},t_{m-})=g_1^{(4)}(\mathbf{z}_m,\mathbf{x},t_{m+})>0,\end{aligned}\right\}\text{for } \mathbf{z}_m\in\partial\Omega_{14}. \tag{9.22}$$

From Luo (2008, 2009a, b, 2012), the criterions of a grazing flow to the relative boundaries of $\partial\Omega_{12}$, $\partial\Omega_{34}$, $\partial\Omega_{23}$ and $\partial\Omega_{14}$ for the controlled slave system are described as

$$\begin{aligned}&\left.\begin{aligned}&G_{\partial\Omega_{12}}^{(0,\alpha)}(\mathbf{z}_m,\mathbf{x},t_{m\pm})=g_2^{(\alpha)}(\mathbf{z}_m,\mathbf{x},t_{m\pm})=0,\\&(-1)^{\alpha}G_{\partial\Omega_{12}}^{(1,\alpha)}(\mathbf{z}_m,\mathbf{x},t_{m\pm})=(-1)^{\alpha}Dg_2^{(\alpha)}(\mathbf{z}_m,\mathbf{x},t_{m\pm})<0\end{aligned}\right\}\\&\text{for } \mathbf{z}_m\in\partial\Omega_{12} \text{ in } \Omega_\alpha(\alpha\in\{1,2\});\\&\left.\begin{aligned}&G_{\partial\Omega_{34}}^{(0,\alpha)}(\mathbf{z}_m,\mathbf{x},t_{m\pm})=g_2^{(\alpha)}(\mathbf{z}_m,\mathbf{x},t_{m\pm})=0,\\&(-1)^{\alpha}G_{\partial\Omega_{34}}^{(1,\alpha)}(\mathbf{z}_m,\mathbf{x},t_{m\pm})=(-1)^{\alpha}Dg_2^{(\alpha)}(\mathbf{z}_m,\mathbf{x},t_{m\pm})>0\end{aligned}\right\}\\&\text{for } \mathbf{z}_m\in\partial\Omega_{34} \text{ in } \Omega_\alpha(\alpha\in\{3,4\});\\&\left.\begin{aligned}&G_{\partial\Omega_{23}}^{(0,\alpha)}(\mathbf{z}_m,\mathbf{x},t_{m\pm})=g_1^{(\alpha)}(\mathbf{z}_m,\mathbf{x},t_{m\pm})=0,\\&(-1)^{\alpha}G_{\partial\Omega_{23}}^{(1,\alpha)}(\mathbf{z}_m,\mathbf{x},t_{m\pm})=(-1)^{\alpha}Dg_1^{(\alpha)}(\mathbf{z}_m,\mathbf{x},t_{m\pm})>0\end{aligned}\right\}\\&\text{for } \mathbf{z}_m\in\partial\Omega_{23} \text{ in } \Omega_\alpha(\alpha\in\{2,3\});\\&\left.\begin{aligned}&G_{\partial\Omega_{14}}^{(0,\alpha)}(\mathbf{z}_m,\mathbf{x},t_{m\pm})=g_1^{(\alpha)}(\mathbf{z}_m,\mathbf{x},t_{m\pm})=0,\\&(-1)^{\alpha}G_{\partial\Omega_{14}}^{(1,\alpha)}(\mathbf{z}_m,\mathbf{x},t_{m\pm})=(-1)^{\alpha}Dg_1^{(\alpha)}(\mathbf{z}_m,\mathbf{x},t_{m\pm})<0\end{aligned}\right\}\\&\text{for } \mathbf{z}_m\in\partial\Omega_{14} \text{ in } \Omega_\alpha(\alpha\in\{1,4\});\end{aligned} \tag{9.23}$$

From Luo (2008, 2009a, b, 2012), the criterions for onset of a sliding flow on the relative boundaries of $\partial\Omega_{12}$, $\partial\Omega_{34}$, $\partial\Omega_{23}$ and $\partial\Omega_{14}$ for the controlled slave system are described as

$$\left.\begin{array}{l} G_{\partial\Omega_{12}}^{(0,1)}(\mathbf{z}_m,\mathbf{x},t_{m-}) = g_2^{(1)}(\mathbf{z}_m,\mathbf{x},t_{m-}) < 0, \\ G_{\partial\Omega_{12}}^{(0,2)}(\mathbf{z}_m,\mathbf{x},t_{m\pm}) = g_2^{(2)}(\mathbf{z}_m,\mathbf{x},t_{m\pm}) = 0, \\ G_{\partial\Omega_{12}}^{(1,2)}(\mathbf{z}_m,\mathbf{x},t_{m\pm}) = Dg_2^{(2)}(\mathbf{z}_m,\mathbf{x},t_{m\pm}) < 0 \end{array}\right\}\text{from } \Omega_1 \to \partial\Omega_{12};$$
$$\left.\begin{array}{l} G_{\partial\Omega_{34}}^{(0,3)}(\mathbf{z}_m,\mathbf{x},t_{m-}) = g_2^{(3)}(\mathbf{z}_m,\mathbf{x},t_{m-}) > 0, \\ G_{\partial\Omega_{34}}^{(0,4)}(\mathbf{z}_m,\mathbf{x},t_{m\pm}) = g_2^{(4)}(\mathbf{z}_m,\mathbf{x},t_{m\pm}) = 0, \\ G_{\partial\Omega_{34}}^{(1,4)}(\mathbf{z}_m,\mathbf{x},t_{m\pm}) = Dg_2^{(4)}(\mathbf{z}_m,\mathbf{x},t_{m\pm}) > 0 \end{array}\right\}\text{from } \Omega_3 \to \partial\Omega_{34};$$
$$\left.\begin{array}{l} G_{\partial\Omega_{23}}^{(0,2)}(\mathbf{z}_m,\mathbf{x},t_{m-}) = g_1^{(2)}(\mathbf{z}_m,\mathbf{x},t_{m-}) < 0, \\ G_{\partial\Omega_{23}}^{(0,3)}(\mathbf{z}_m,\mathbf{x},t_{m\pm}) = g_1^{(3)}(\mathbf{z}_m,\mathbf{x},t_{m\pm}) = 0, \\ G_{\partial\Omega_{23}}^{(1,3)}(\mathbf{z}_m,\mathbf{x},t_{m\pm}) = Dg_1^{(3)}(\mathbf{z}_m,\mathbf{x},t_{m\pm}) < 0 \end{array}\right\}\text{from } \Omega_2 \to \partial\Omega_{23};$$
$$\left.\begin{array}{l} G_{\partial\Omega_{14}}^{(0,4)}(\mathbf{z}_m,\mathbf{x},t_{m-}) = g_1^{(4)}(\mathbf{z}_m,\mathbf{x},t_{m-}) > 0, \\ G_{\partial\Omega_{14}}^{(0,1)}(\mathbf{z}_m,\mathbf{x},t_{m\pm}) = g_1^{(1)}(\mathbf{z}_m,\mathbf{x},t_{m\pm}) = 0, \\ G_{\partial\Omega_{14}}^{(1,1)}(\mathbf{z}_m,\mathbf{x},t_{m\pm}) = Dg_1^{(1)}(\mathbf{z}_m,\mathbf{x},t_{m\pm}) > 0 \end{array}\right\}\text{from } \Omega_4 \to \partial\Omega_{14}. \tag{9.24}$$

From Luo (2008, 2009a, b, 2012), the vanishing criterions of a sliding flow from $\partial\Omega_{12}$, $\partial\Omega_{34}$, $\partial\Omega_{23}$ and $\partial\Omega_{14}$ to a domain for the controlled slave system are written as ($\beta \neq \alpha$)

$$\begin{array}{l} \left.\begin{array}{l} (-1)^{\beta} G_{\partial\Omega_{12}}^{(0,\beta)}(\mathbf{z}_m,\mathbf{x},t_{m-}) = (-1)^{\beta} g_2^{(\beta)}(\mathbf{z}_m,\mathbf{x},t_{m-}) > 0 \\ G_{\partial\Omega_{12}}^{(0,\alpha)}(\mathbf{z}_m,\mathbf{x},t_{m\mp}) = g_2^{(\alpha)}(\mathbf{z}_m,\mathbf{x},t_{m\mp}) = 0 \\ (-1)^{\alpha} G_{\partial\Omega_{12}}^{(1,\alpha)}(\mathbf{z}_m,\mathbf{x},t_{m\mp}) = (-1)^{\alpha} Dg_2^{(\alpha)}(\mathbf{z}_m,\mathbf{x},t_{m\mp}) < 0, \end{array}\right\}\text{from } \partial\Omega_{12} \to \Omega_{\alpha}, \\ \text{for } \mathbf{z}_m \in \partial\Omega_{12}, \text{with } \alpha,\beta \in \{1,2\}; \\ \left.\begin{array}{l} (-1)^{\beta} G_{\partial\Omega_{34}}^{(0,\beta)}(\mathbf{z}_m,\mathbf{x},t_{m-}) = (-1)^{\beta} g_2^{(\beta)}(\mathbf{z}_m,\mathbf{x},t_{m-}) < 0 \\ G_{\partial\Omega_{34}}^{(0,\alpha)}(\mathbf{z}_m,\mathbf{x},t_{m\mp}) = g_2^{(\alpha)}(\mathbf{z}_m,\mathbf{x},t_{m\mp}) = 0 \\ (-1)^{\alpha} G_{\partial\Omega_{34}}^{(1,\alpha)}(\mathbf{z}_m,\mathbf{x},t_{m\mp}) = (-1)^{\alpha} Dg_2^{(\alpha)}(\mathbf{z}_m,\mathbf{x},t_{m\mp}) > 0, \end{array}\right\}\text{from } \partial\Omega_{34} \to \Omega_{\alpha}, \\ \text{for } \mathbf{z}_m \in \partial\Omega_{34}, \text{with } \alpha,\beta \in \{3,4\}; \end{array} \tag{9.25a}$$

$$\left.\begin{array}{l}(-1)^{\beta}G^{(0,\beta)}_{\partial\Omega_{23}}(\mathbf{z}_m,\mathbf{x},t_{m-})=(-1)^{\beta}g_1^{(\beta)}(\mathbf{z}_m,\mathbf{x},t_{m-})<0\\ G^{(0,\alpha)}_{\partial\Omega_{23}}(\mathbf{z}_m,\mathbf{x},t_{m\mp})=g_1^{(\alpha)}(\mathbf{z}_m,\mathbf{x},t_{m\mp})=0\\ (-1)^{\alpha}G^{(1,\alpha)}_{\partial\Omega_{23}}(\mathbf{z}_m,\mathbf{x},t_{m\mp})=(-1)^{\alpha}Dg_1^{(\alpha)}(\mathbf{z}_m,\mathbf{x},t_{m\mp})>0,\end{array}\right\}\text{from }\partial\Omega_{23}\to\Omega_{\alpha},$$
$$\text{for }\mathbf{z}_m\in\partial\Omega_{23},\text{ with }\alpha,\beta\in\{2,3\};$$
$$\left.\begin{array}{l}(-1)^{\beta}G^{(0,\beta)}_{\partial\Omega_{14}}(\mathbf{z}_m,\mathbf{x},t_{m-})=(-1)^{\beta}g_1^{(\beta)}(\mathbf{z}_m,\mathbf{x},t_{m-})>0\\ G^{(0,\alpha)}_{\partial\Omega_{14}}(\mathbf{z}_m,\mathbf{x},t_{m\mp})=g_1^{(\alpha)}(\mathbf{z}_m,\mathbf{x},t_{m\mp})=0\\ (-1)^{\alpha}G^{(1,\alpha)}_{\partial\Omega_{14}}(\mathbf{z}_m,\mathbf{x},t_{m\mp})=(-1)^{\alpha}Dg_1^{(\alpha)}(\mathbf{z}_m,\mathbf{x},t_{m\mp})<0,\end{array}\right\}\text{from }\partial\Omega_{14}\to\Omega_{\alpha},$$
$$\text{for }\mathbf{z}_m\in\partial\Omega_{14},\text{ with }\alpha,\beta\in\{1,4\};\tag{9.25b}$$

9.3 Switching Projective

From Luo (2008, 2009a, b, 2012), the functional synchronization of the controlled Duffing system with two master systems occurs at the edge of all the boundaries under the switching controller in Eq. 9.5. The corresponding synchronization conditions are described as

$$\begin{array}{l}\left.\begin{array}{l}G^{(1)}_{\partial\Omega_{14}}(\mathbf{z}_m,\mathbf{x},t_{m-})=g_1^{(1)}(\mathbf{z}_m,\mathbf{x},t_{m-})<0,\\ G^{(1)}_{\partial\Omega_{12}}(\mathbf{z}_m,\mathbf{x},t_{m-})=g_2^{(1)}(\mathbf{z}_m,\mathbf{x},t_{m-})<0\end{array}\right\}\text{for }\mathbf{z}_m\in\partial\Omega_{12}\cap\partial\Omega_{14}\text{ on }\Omega_1;\\ \left.\begin{array}{l}G^{(2)}_{\partial\Omega_{12}}(\mathbf{z}_m,\mathbf{x},t_{m-})=g_2^{(2)}(\mathbf{z}_m,\mathbf{x},t_{m-})>0,\\ G^{(2)}_{\partial\Omega_{23}}(\mathbf{z}_m,\mathbf{x},t_{m-})=g_1^{(2)}(\mathbf{z}_m,\mathbf{x},t_{m-})<0\end{array}\right\}\text{for }\mathbf{z}_m\in\partial\Omega_{12}\cap\partial\Omega_{23}\text{ on }\Omega_2;\\ \left.\begin{array}{l}G^{(3)}_{\partial\Omega_{23}}(\mathbf{z}_m,\mathbf{x},t_{m-})=g_1^{(3)}(\mathbf{z}_m,\mathbf{x},t_{m-})>0,\\ G^{(3)}_{\partial\Omega_{34}}(\mathbf{z}_m,\mathbf{x},t_{m-})=g_2^{(3)}(\mathbf{z}_m,\mathbf{x},t_{m-})>0\end{array}\right\}\text{for }\mathbf{z}_m\in\partial\Omega_{23}\cap\partial\Omega_{34}\text{ on }\Omega_3;\\ \left.\begin{array}{l}G^{(4)}_{\partial\Omega_{34}}(\mathbf{z}_m,\mathbf{x},t_{m-})=g_2^{(4)}(\mathbf{z}_m,\mathbf{x},t_{m-})<0,\\ G^{(4)}_{\partial\Omega_{14}}(\mathbf{z}_m,\mathbf{x},t_{m-})=g_1^{(4)}(\mathbf{z}_m,\mathbf{x},t_{m-})>0\end{array}\right\}\text{for }\mathbf{z}_m\in\partial\Omega_{23}\cap\partial\Omega_{34}\text{ on }\Omega_4.\end{array}\tag{9.26}$$

From the controlled vector fields and the coordinate transformation in 9.10, the basic functions can thus be rewritten as

$$
\begin{aligned}
&g_1\left(\mathbf{z}^{(\alpha)},\mathbf{x},t\right) \equiv g_1^{(1)}\left(\mathbf{z}^{(\alpha)},\mathbf{x},t\right) = z_2^{(\alpha)} - k_1 \text{ in } \Omega_\alpha \text{ for } \alpha = 1,2;\\
&g_2\left(\mathbf{z}^{(\alpha)},\mathbf{x},t\right) \equiv g_1^{(1)}\left(\mathbf{z}^{(\alpha)},\mathbf{x}\,t\right) = z_2^{(\alpha)} + k_1 \text{ in } \Omega_\alpha \text{ for } \alpha = 3,4;\\
&g_3\left(\mathbf{z}^{(\alpha)},\mathbf{x},t\right) \equiv g_2^{(1)}\left(\mathbf{z}^{(\alpha)},\mathbf{x},t\right) = G_w\left(\mathbf{z}^{(\alpha)},\mathbf{x},t\right) - k_2 + (p_1x_2+p_2y_2)^2\\
&\quad \sin(p_1x_1+p_2y_1) - \left[p_1\left(x_2\left(1-x_1^2\right)b - x_1 + Q\sin(\Omega t)\right)+\right.\\
&\quad \left.p_2(-\alpha_0\sin y_1 + Q_0\cos\Omega_0 t)\right]\cos(p_1x_1+p_2y_1) \text{ in } \Omega_\alpha \text{ for } \alpha = 1,4;\\
&g_4\left(\mathbf{z}^{(\alpha)},\mathbf{x},t\right) \equiv g_2^{(1)}\left(\mathbf{z}^{(\alpha)},\mathbf{x},t\right) = G_w\left(\mathbf{z}^{(\alpha)},\mathbf{x},t\right) + k_2 + (p_1x_2+p_2y_2)^2\\
&\quad \sin(p_1x_1+p_2y_1) - \left[p_1\left(x_2\left(1-x_1^2\right)b - x_1 + Q\sin(\Omega t)\right)+\right.\\
&\quad \left.p_2(-\alpha_0\sin y_1 + Q_0\cos\Omega_0 t)\right]\cos(p_1x_1+p_2y_1) \text{ in } \Omega_\alpha \text{ for } \alpha = 2,3.
\end{aligned}
\tag{9.27}
$$

where

$$
\begin{aligned}
G_w\left(\mathbf{z}^{(\alpha)},\mathbf{x},t\right) = &\left(-\delta\left(z_2^{(\alpha)} + (p_1x_2+p_2y_2)\cos(p_1x_1+p_2y_1)\right)\right.\\
&+\alpha_1\left(z_1^{(\alpha)} + \sin(p_1x_1+p_2y_1)\right)\\
&\left.-\alpha_2\left(z_1^{(\alpha)} + \sin(p_1x_1+p_2y_1)\right)^3 + Q_1\cos\Omega_1 t\right).
\end{aligned}
\tag{9.28}
$$

The switching synchronization conditions of three systems under specific non-smooth control laws in Eq. 9.5 are shown as

$$
\begin{aligned}
&g_1(\mathbf{z}_m,\mathbf{x},t_{m-}) = z_{2m} - k_1 < 0,\\
&g_2(\mathbf{z}_m,\mathbf{x},t_{m-}) = z_{2m} + k_1 > 0,\\
&g_3(\mathbf{z}_m,\mathbf{x},t_{m-}) = G_w\left(\mathbf{z}^{(\alpha)},\mathbf{x},t\right) - k_2 + (p_1x_2+p_2y_2)^2\\
&\qquad \sin(p_1x_1+p_2y_1) - \left[p_1\left(x_2\left(1-x_1^2\right)b - x_1 + Q\sin(\Omega t)\right)+\right.\\
&\qquad \left.p_2(-\alpha_0\sin y_1 + Q_0\cos\Omega_0 t)\right]\cos(p_1x_1+p_2y_1) < 0,\\
&g_4(\mathbf{z}_m,\mathbf{x},t_{m-}) = G_w\left(\mathbf{z}^{(\alpha)},\mathbf{x},t\right) + k_2 + (p_1x_2+p_2y_2)^2\\
&\qquad \sin(p_1x_1+p_2y_1) - \left[p_1\left(x_2\left(1-x_1^2\right)b - x_1 + Q\sin(\Omega t)\right)+\right.\\
&\qquad \left.p_2(-\alpha_0\sin y_1 + Q_0\cos\Omega_0 t)\right]\cos(p_1x_1+p_2y_1) > 0.
\end{aligned}
\tag{9.29}
$$

Let $\mathbf{z}_m = 0$—the coexisting conditions of switching combination synchronization of controlled Duffing system with two master systems are

$$
\begin{aligned}
&g_1(\mathbf{z}_m,\mathbf{x},t_{m-}) = -k_1 < 0,\\
&g_2(\mathbf{z}_m,\mathbf{x},t_{m-}) = +k_1 > 0,\\
&g_3(\mathbf{z}_m,\mathbf{x},t_{m-}) = G_w(\mathbf{x},t_{m-}) - k_2 + (p_1x_2+p_2y_2)^2\\
&\qquad \sin(p_1x_1+p_2y_1) - \left[p_1\left(x_2\left(1-x_1^2\right)b - x_1 + Q\sin(\Omega t)\right)+\right.\\
&\qquad \left.p_2(-\alpha_0\sin y_1 + Q_0\cos\Omega_0 t)\right]\cos(p_1x_1+p_2y_1) < 0,\\
&g_4(\mathbf{z}_m,\mathbf{x},t_{m-}) = G_w(\mathbf{x},t_{m-}) + k_2 + (p_1x_2+p_2y_2)^2\\
&\qquad \sin(p_1x_1+p_2y_1) - \left[p_1\left(x_2\left(1-x_1^2\right)b - x_1 + Q\sin(\Omega t)\right)+\right.\\
&\qquad \left.p_2(-\alpha_0\sin y_1 + Q_0\cos\Omega_0 t)\right]\cos(p_1x_1+p_2y_1) > 0.
\end{aligned}
\tag{9.30}
$$

where

$$
\begin{aligned}
G_w(\mathbf{x}, t_{m-}) = & (-\delta(p_1x_2 + p_2y_2)\cos(p_1x_1 + p_2y_1)) \\
& + \alpha_1(\sin(p_1x_1 + p_2y_1)) \\
& - \alpha_2(\sin(p_1x_1 + p_2y_1))^3 + Q_1\cos\Omega_1 t).
\end{aligned} \tag{9.31}
$$

With $k_1 > 0$ and $k_2 > 0$, the first two equations of Eq. 9.30 can be automatically satisfied, and other equations provide the set of synchronous invariant as

$$
\begin{aligned}
-k_2 < G_w(\mathbf{x}, t_{m-}) + (p_1x_2 + p_2y_2)^2 \sin(p_1x_1 + p_2y_1) - \left[p_1\left(x_2\left(1 - x_1^2\right)b - x_1\right.\right. \\
\left.\left. + Q\sin(\Omega t)\right) + p_2(-\alpha_0 \sin y_1 + Q_0\cos\Omega_0 t)\right]\cos(p_1x_1 + p_2y_1) < k_2
\end{aligned} \tag{9.32}
$$

In a very small neighborhood of the switching function synchronization for three systems of $\mathbf{z}_m = 0$, the attractivity synchronization conditions are provided, for $|\mathbf{z} - \mathbf{z}_m| < \varepsilon$, as follows:

$$
\begin{aligned}
& 0 \le z_2 < k_1 \text{ and } G_w(\mathbf{x}, t_{m-}) + (p_1x_2 + p_2y_2)^2 \sin(p_1x_1 + p_2y_1) - \\
& \quad \left[p_1\left(x_2\left(1 - x_1^2\right)b - x_1 + Q\sin(\Omega t)\right) + p_2(-\alpha_0 \sin y_1 + Q_0\cos\Omega_0 t)\right] \\
& \quad \cos(p_1x_1 + p_2y_1) < k_2 \qquad \text{for } z_1 \in [0, \infty) \text{ in } \Omega_1, \\
& 0 \le z_2 < k_1 \text{ and } -k_2 < G_w(\mathbf{x}, t_{m-}) + (p_1x_2 + p_2y_2)^2 \sin(p_1x_1 + p_2y_1) - \\
& \quad \left[p_1\left(x_2\left(1 - x_1^2\right)b - x_1 + Q\sin(\Omega t)\right) + p_2(-\alpha_0 \sin y_1 + Q_0\cos\Omega_0 t)\right] \\
& \quad \cos(p_1x_1 + p_2y_1) \qquad \text{for } z_1 \in [0, \infty) \text{ in } \Omega_2, \\
& -k_1 < z_2 \le 0 \text{ and } -k_2 < G_w(\mathbf{x}, t_{m-}) + (p_1x_2 + p_2y_2)^2 \sin(p_1x_1 + p_2y_1) - \\
& \quad \left[p_1\left(x_2\left(1 - x_1^2\right)b - x_1 + Q\sin(\Omega t)\right) + p_2(-\alpha_0 \sin y_1 + Q_0\cos\Omega_0 t)\right] \\
& \quad \cos(p_1x_1 + p_2y_1) \qquad \text{for } z_1 \in (-\infty, 0] \text{ in } \Omega_3, \\
& -k_1 < z_2 \le 0 \text{ and } G_w(\mathbf{x}, t_{m-}) + (p_1x_2 + p_2y_2)^2 \sin(p_1x_1 + p_2y_1) - \\
& \quad \left[p_1\left(x_2\left(1 - x_1^2\right)b - x_1 + Q\sin(\Omega t)\right) + p_2(-\alpha_0 \sin y_1 + Q_0\cos\Omega_0 t)\right] \\
& \quad \cos(p_1x_1 + p_2y_1) < k_2 \qquad \text{for } z_1 \in (-\infty, 0] \text{ in } \Omega_4.
\end{aligned} \tag{9.33}
$$

Suppose that z_1^*, z_2^*, z_3^* and z_4^* are known, the initial values for the chaotic synchronization of three systems are determined by

$$
\begin{aligned}
w_1 &= z_1^* + \sin(p_1x_1 + p_2y_1) \text{ and} \\
w_2 &= z_2^* + (p_1x_2 + p_2y_2)\cos(p_1x_1 + p_2y_1).
\end{aligned} \tag{9.34}
$$

From a sliding flow disappearing on the relative boundary, the vanishing criterions for switching synchronization of three chaotic systems for $\mathbf{z}^{(\alpha)}(t_{m\mp}) = \mathbf{z}_m^{(\alpha)} = \mathbf{z}_m$ are written as

$$\text{For } (\alpha,\beta)=\{(1,4),(2,3)\},$$
$$\left.\begin{array}{l} g_1\left(\mathbf{z}_m^{(\alpha)},\mathbf{x},t_{m\mp}\right)=z_{2m}^{(\alpha)}-k_1=0,\\ Dg_1\left(\mathbf{z}_m^{(\alpha)},\mathbf{x},t_{m\mp}\right)>0,\\ g_2\left(\mathbf{z}_m^{(\beta)},\mathbf{x},t_{m-}\right)=z_{2m}^{(\beta)}+k_1>0 \end{array}\right\}$$
$$\text{from } z_{m+\varepsilon}=w_1-\sin(p_1x_1+p_2y_1)>0; \tag{9.35}$$
$$\left.\begin{array}{l} g_1\left(\mathbf{z}_m^{(\alpha)},\mathbf{x},t_{m-}\right)=z_{2m}^{(\alpha)}-k_1<0,\\ g_2\left(\mathbf{z}_m^{(\beta)},\mathbf{x},t_{m\mp}\right)=z_{2m}^{(\beta)}+k_1=0,\\ Dg_2\left(\mathbf{z}_m^{(\beta)},\mathbf{x},t_{m\mp}\right)<0 \end{array}\right\}$$
$$\text{from } z_{m+\varepsilon}=w_1-\sin(p_1x_1+p_2y_1)<0;$$

$$\text{For } (\alpha,\beta)=\{(1,2),(4,3)\},$$
$$\left.\begin{array}{l} g_3\left(\mathbf{z}_m^{(\alpha)},\mathbf{x},t_{m\mp}\right)=G_w\left(\mathbf{z}^{(\alpha)},\mathbf{x},t_{m-}\right)-k_2+(p_1x_2+p_2y_2)^2\\ \quad\sin(p_1x_1+p_2y_1)-\left[p_1\left(x_2\left(1-x_1^2\right)b-x_1+Q\sin(\Omega t)\right)\right.\\ \quad\left.+p_2(-\alpha_0\sin y_1+Q_0\cos\Omega_0t)\right]\cos(p_1x_1+p_2y_1)=0,\\ Dg_3\left(\mathbf{z}_m^{(\alpha)},\mathbf{x},t_{m\mp}\right)>0,\\ g_4\left(\mathbf{z}_m^{(\beta)},\mathbf{x},t_{m-}\right)=G_w\left(\mathbf{z}^{(\alpha)},\mathbf{x},t_{m-}\right)+k_2+(p_1x_2+p_2y_2)^2\\ \quad\sin(p_1x_1+p_2y_1)-\left[p_1\left(x_2\left(1-x_1^2\right)b-x_1+Q\sin(\Omega t)\right)\right.\\ \quad\left.+p_2(-\alpha_0\sin y_1+Q_0\cos\Omega_0t)\right]\cos(p_1x_1+p_2y_1)>0 \end{array}\right\}$$
$$\text{from } z_{m+\varepsilon}=w_2-(p_1x_2+p_2y_2)\cos(p_1x_1+p_2y_1)>0; \tag{9.36}$$
$$\left.\begin{array}{l} g_3\left(\mathbf{z}_m^{(\alpha)},\mathbf{x},t_{m\mp}\right)=G_w\left(\mathbf{z}^{(\alpha)},\mathbf{x},t_{m-}\right)-k_2+(p_1x_2+p_2y_2)^2\\ \quad\sin(p_1x_1+p_2y_1)-\left[p_1\left(x_2\left(1-x_1^2\right)b-x_1+Q\sin(\Omega t)\right)\right.\\ \quad\left.+p_2(-\alpha_0\sin y_1+Q_0\cos\Omega_0t)\right]\cos(p_1x_1+p_2y_1)<0,\\ g_4\left(\mathbf{z}_m^{(\beta)},\mathbf{x},t_{m-}\right)=G_w\left(\mathbf{z}^{(\alpha)},\mathbf{x},t_{m-}\right)+k_2+(p_1x_2+p_2y_2)^2\\ \quad\sin(p_1x_1+p_2y_1)-\left[p_1\left(x_2\left(1-x_1^2\right)b-x_1+Q\sin(\Omega t)\right)\right.\\ \quad\left.+p_2(-\alpha_0\sin y_1+Q_0\cos\Omega_0t)\right]\cos(p_1x_1+p_2y_1)=0,\\ Dg_4\left(\mathbf{z}_m^{(\beta)},\mathbf{x},t_{m\mp}\right)<0 \end{array}\right\}$$
$$\text{from } z_{m+\varepsilon}=w_2-(p_1x_2+p_2y_2)\cos(p_1x_1+p_2y_1)<0.$$

From the appearance of sliding flow on the relative boundary, the onset criterions for function projective synchronization of three systems for $\mathbf{z}^{(\alpha)}(t_{m\mp})=\mathbf{z}_m^{(\alpha)}=\mathbf{z}_m$ are given as follows:

$$\text{For } (\alpha,\beta)=\{(1,4),(2,3)\},$$

$$\left.\begin{aligned}
&g_1\left(\mathbf{z}_m^{(\alpha)},\mathbf{x},t_{m\mp}\right)=z_{2m}^{(\alpha)}-k_1=0,\\
&Dg_1\left(\mathbf{z}_m^{(\alpha)},\mathbf{x},t_{m\mp}\right)>0,\\
&g_2\left(\mathbf{z}_m^{(\beta)},\mathbf{x},t_{m-}\right)=z_{2m}^{(\beta)}+k_1>0
\end{aligned}\right\}$$
$$\begin{aligned}
&\text{from } z_{m+\varepsilon}=w_1-\sin(p_1x_1+p_2y_1)>0,\\
&\left.\begin{aligned}
&g_1\left(\mathbf{z}_m^{(\alpha)},\mathbf{x},t_{m-}\right)=z_{2m}^{(\alpha)}-k_1<0,\\
&g_2\left(\mathbf{z}_m^{(\beta)},\mathbf{x},t_{m\mp}\right)=z_{2m}^{(\beta)}+k_1=0,\\
&Dg_2\left(\mathbf{z}_m^{(\beta)},\mathbf{x},t_{m\mp}\right)<0
\end{aligned}\right\}\\
&\text{from } z_{m+\varepsilon}=w_1-\sin(p_1x_1+p_2y_1)<0;
\end{aligned} \tag{9.37}$$

For $(\alpha,\beta)=\{(1,2),(4,3)\}$,

$$\begin{aligned}
&\left.\begin{aligned}
&g_3\left(\mathbf{z}_m^{(\alpha)},\mathbf{x},t_{m\mp}\right)=G_w\left(\mathbf{z}^{(\alpha)},\mathbf{x},t_{m-}\right)-k_2+(p_1x_2+p_2y_2)^2\\
&\quad\sin(p_1x_1+p_2y_1)-\left[p_1\left(x_2\left(1-x_1^2\right)b-x_1+Q\sin(\Omega t)\right)\right.\\
&\quad\left.+p_2(-\alpha_0\sin y_1+Q_0\cos\Omega_0 t)\right]\cos(p_1x_1+p_2y_1)=0,\\
&Dg_3\left(\mathbf{z}_m^{(\alpha)},\mathbf{x},t_{m\mp}\right)>0,\\
&g_4\left(\mathbf{z}_m^{(\beta)},\mathbf{x},t_{m-}\right)=G_w\left(\mathbf{z}^{(\alpha)},\mathbf{x},t_{m-}\right)+k_2+(p_1x_2+p_2y_2)^2\\
&\quad\sin(p_1x_1+p_2y_1)-\left[p_1\left(x_2\left(1-x_1^2\right)b-x_1+Q\sin(\Omega t)\right)\right.\\
&\quad\left.+p_2(-\alpha_0\sin y_1+Q_0\cos\Omega_0 t)\right]\cos(p_1x_1+p_2y_1)>0
\end{aligned}\right\}\\
&\text{from } z_{m+\varepsilon}=w_2-(p_1x_2+p_2y_2)\cos(p_1x_1+p_2y_1)>0;\\
&\left.\begin{aligned}
&g_3\left(\mathbf{z}_m^{(\alpha)},\mathbf{x},t_{m\mp}\right)=G_w\left(\mathbf{z}^{(\alpha)},\mathbf{x},t_{m-}\right)-k_2+(p_1x_2+p_2y_2)^2\\
&\quad\sin(p_1x_1+p_2y_1)-\left[p_1\left(x_2\left(1-x_1^2\right)b-x_1+Q\sin(\Omega t)\right)\right.\\
&\quad\left.+p_2(-\alpha_0\sin y_1+Q_0\cos\Omega_0 t)\right]\cos(p_1x_1+p_2y_1)<0,\\
&g_4\left(\mathbf{z}_m^{(\beta)},\mathbf{x},t_{m-}\right)=G_w\left(\mathbf{z}^{(\alpha)},\mathbf{x},t_{m-}\right)+k_2+(p_1x_2+p_2y_2)^2\\
&\quad\sin(p_1x_1+p_2y_1)-\left[p_1\left(x_2\left(1-x_1^2\right)b-x_1+Q\sin(\Omega t)\right)\right.\\
&\quad\left.+p_2(-\alpha_0\sin y_1+Q_0\cos\Omega_0 t)\right]\cos(p_1x_1+p_2y_1)=0,\\
&Dg_4\left(\mathbf{z}_m^{(\beta)},\mathbf{x},t_{m\mp}\right)<0
\end{aligned}\right\}\\
&\text{from } z_{m+\varepsilon}=w_2-(p_1x_2+p_2y_2)\cos(p_1x_1+p_2y_1)<0.
\end{aligned} \tag{9.38}$$

9.4 Synchronization Scenario and Parameter Maps

Consider dynamical system parameters as follows:

$$\begin{aligned}
&\text{Van der Pol system}: b=3, Q=5, \Omega=1.788,\\
&\text{Pendulum system}: \alpha_0=1, Q_0=0.275, \Omega_0=2.18519,\\
&\text{Duffing oscillator}: \alpha_1=1, \alpha_2=1, Q_1=0.4, \Omega_1=1, \delta=0.25,\\
&\text{Projective coefficients}: p_1=0.4, p_2=0.6.
\end{aligned} \tag{9.39}$$

Suppose $k_1 = 5$, according to the analytical conditions in Eqs. 9.35–9.38, the combination synchronization switching scenarios of the controlled slave system (i.e., the Duffing system) varying with k_2 can be calculated and plotted as shown in Fig. 9.2, where "NS" represents "nonsynchronization," "PS" means "partial synchronization," and "FS" are "full synchronizations"; the black dots represent the initial points of the system synchronization; and the red represents the synchronous vanishing points.

In Fig. 9.2, the switching combination synchronization of three systems cannot be observed for $k_2 < 0.0$. The partial switching synchronization of the controlled Duffing system with van der Pol and pendulum lies in $k_2 \in (0, 16.6913)$. Three distinct systems enter full synchronization for $k_2 > 16.6913$ without vanishing points observed. Also in Fig. 9.2, it can be seen that k_2 must be positive to obtain partial combined synchronization. With increasing k_2, the synchronization switching points become less and less until the systems realize full synchronization, which means that the larger the k_2, the easier it is to realize the full switching combination synchronization.

For a global view of switching combination synchronization of three different nonautonomous systems, the parameter maps are calculated by changing k_1 and k_2 , as shown in Fig. 9.3, where partial synchronization area is labeled with shade part. It can be intuitively observed from Fig. 9.3, for $k_2 < 0$, that no matter how k_1 changes, the controlled Duffing systems cannot be switching synchronized with the van der Pol and pendulum under the control constraints. For $k_2 \in (0, 16.6913)$, the partial combined synchronization exists. For $k_2 > 16.6913$, the full combined synchronization is observed.

9.5 Numerical Illustration

The partial combination synchronization with $k_1 = 5$, $k_2 = 3$ and the full combination synchronization with $k_1 = 5$, $k_2 = 20$ are also calculated as shown in Figs. 9.4 and 9.5, where all hollow white circles stand for the appearance of the combination synchronization, solid green circles for the vanishing of the combination synchronization, and the gray for synchronization areas abbreviated as "S" and nonsynchronization areas as "N."

In Fig. 9.4a–d, the time histories of state variables of two master systems and slave system, the errors of synchronization, and G - functions are presented, respectively. The shaded and white regions denote the combination synchronization and nonsynchronization, respectively. In Fig. 9.4a, the state variables x_1, y_1 and w_1 are synchronized in the synchronization parts, which are satisfied with the relation $w_1 = \sin(p_1 x_1 + p_2 y_1)$, while in nonsynchronization regions, the three state variables cannot obtain the combination synchronization. In Fig. 9.4b, the three state variables x_2, y_2 and w_2 are synchronized with $w_2 = (p_1 x_2 + p_2 y_2)\cos(p_1 x_1 + p_2 y_1)$. In Fig. 9.4c, the time histories of the errors z_1 and z_2 are depicted, where the combination synchronization of the three distinct system possesses $z_1 = 0.0$

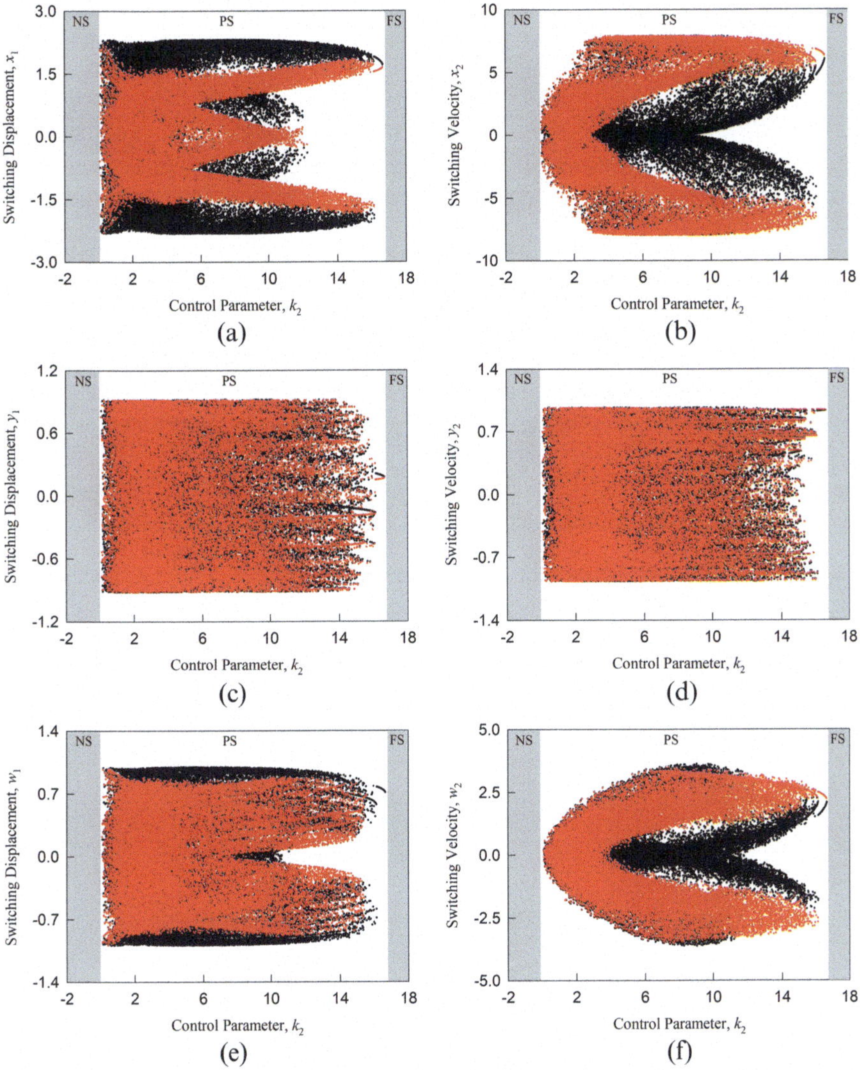

Fig. 9.2 Synchronization scenario of the switching points versus control parameter k_2: (**a**) switching displacement and (**b**) switching velocity of van der Pol, (**c**) switching displacement and (**d**) switching velocity of pendulum, (**e**) switching displacement and (**f**) switching velocity of Duffing, (**g**) switching phase, (**h**) switching points in phase space of van der Pol, (**i**) switching points in phase space of pendulum, (**j**) switching points in phase space of Duffing (control parameter: $k_1 = 5$; initial condition:$x_1 = 0.2$, $x_2 = 0.3$, $y_1 = 0.166622996043$, $y_2 = 0.784536254315$,$w_1 = \sin(p_1{}^*x_1, +p_2{}^*y_1)$, $w_2 = (p_1{}^*x_2 + p_2{}^*y_2) * \cos(p_1{}^*x_1 + p_2{}^*y_1)$)

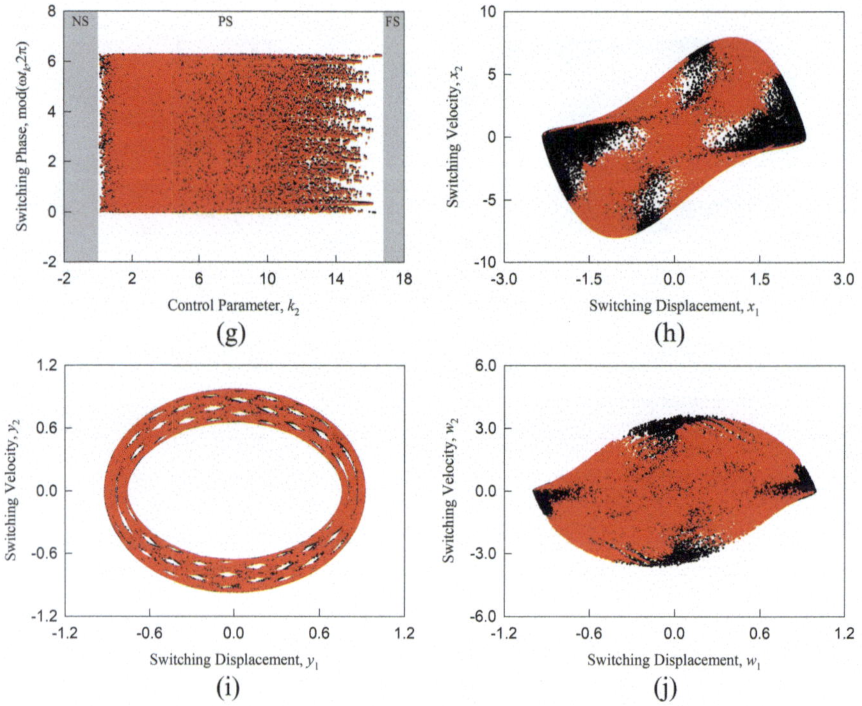

Fig. 9.2 (continued)

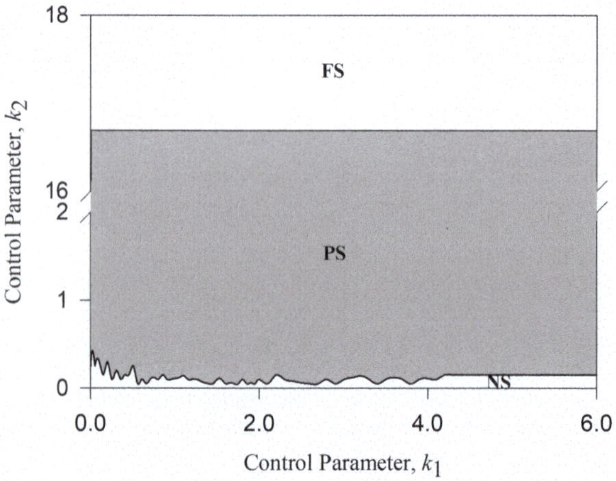

Fig. 9.3 Control parameter map of (k_1, k_2) for the synchronicity of three systems. (Initial condition: $x_1 = 0.2$, $x_2 = 0.3$, $y_1 = 0.166622996043$, $y_2 = 0.784536254315$, $w_1 = \sin(p_1{}^{*}x_1 + p_2{}^{*}y_1)$, $w_2 = (p_1{}^{*}x_2 + p_2{}^{*}y_2) * \cos(p_1{}^{*}x_1 + p_2{}^{*}y_1)$)

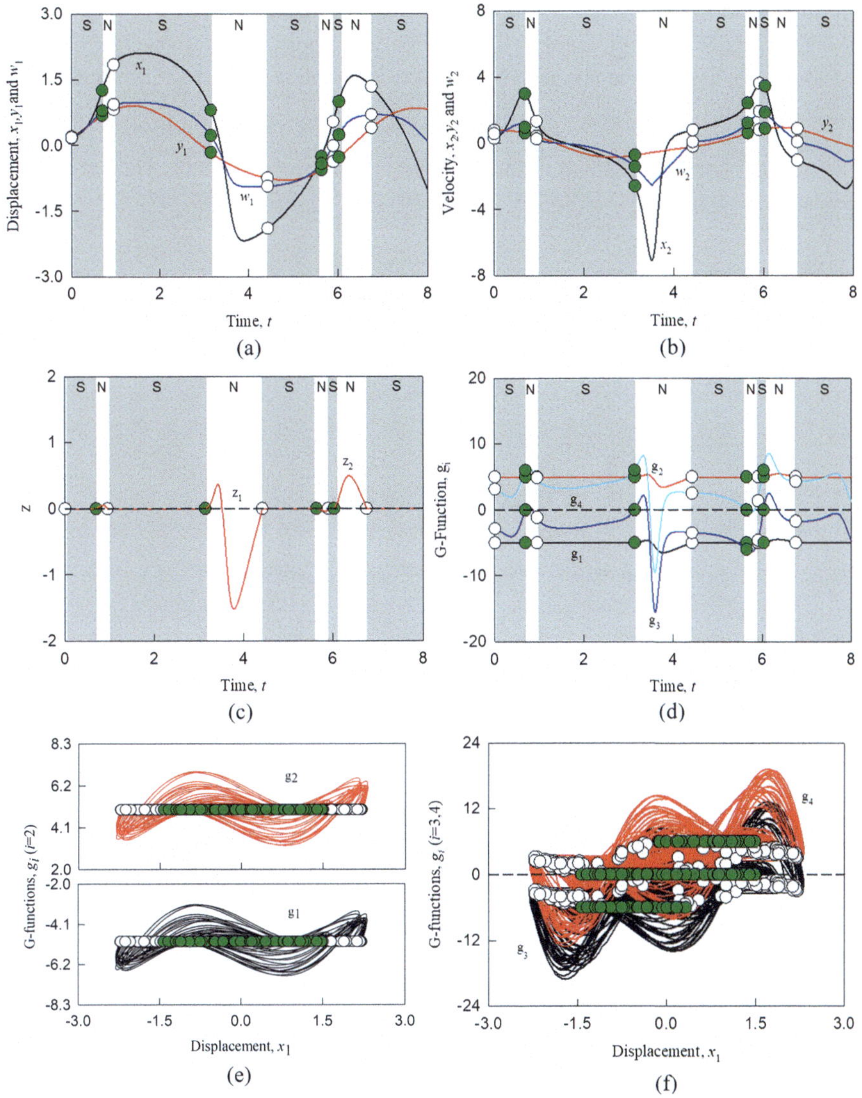

Fig. 9.4 The partial synchronization for three systems: (**a**) displacement responses, (**b**) velocity responses, (**c**) error responses, (**d**) G-functions, (**e**) G-function ($g_{1,\ 2}$) verse displacement x_1, (**f**) G-function ($g_{3,\ 4}$) verse displacement x_1. (Control parameters: $k_1 = 5$, $k_2 = 3$)

and $z_2 = 0.0$, and the nonsynchronization cannot make the errors stable. In Fig. 9.4d, the time history G-functions $g_i(i = 1, 2, 3, 4)$ are presented to illustrate the conditions of combination synchronization. The G-functions $g_i(i = 1, 2)$ always hold the conditions $g_1 < 0$ and $g_2 > 0$, and the G-functions $g_i(i = 3, 4)$ are satisfied with $g_3 < 0$ and $g_4 > 0$ for the combination synchronization, while

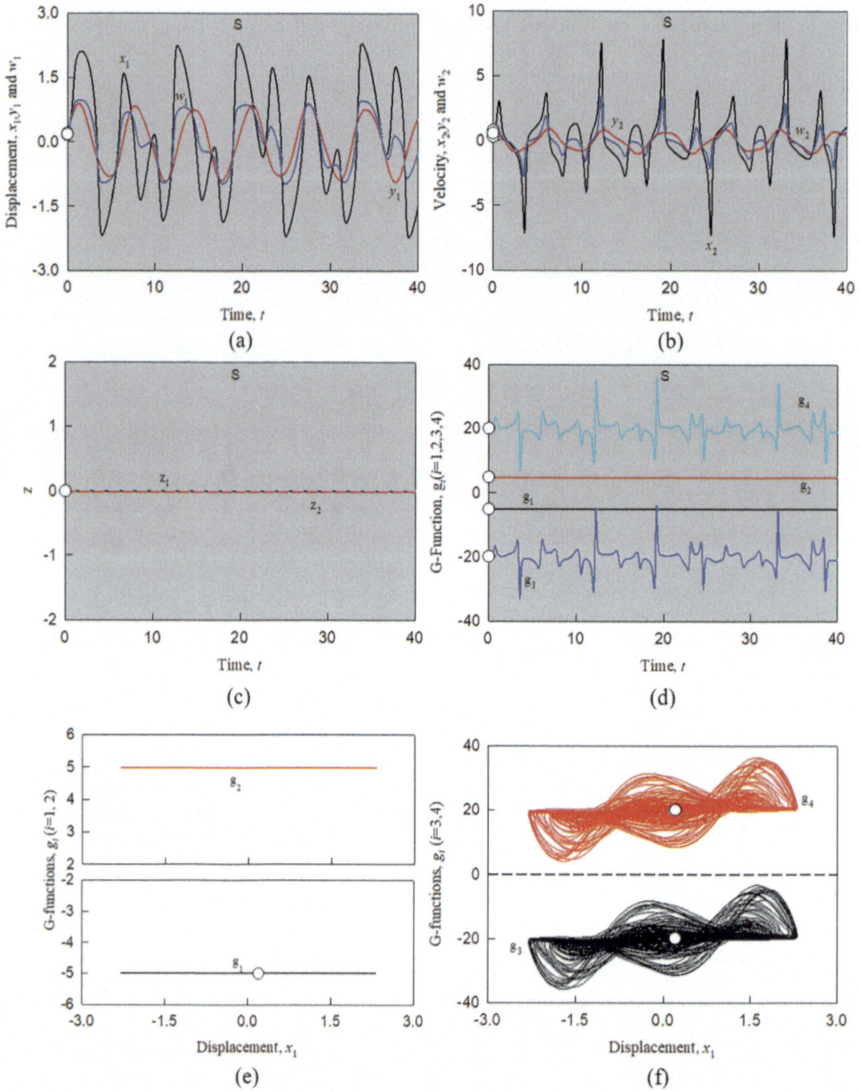

Fig. 9.5 The full synchronization for three systems: (**a**) displacement responses, (**b**) velocity responses, (**c**) error responses, (**d**) G-functions, (**e**) G-function ($g_{1,\ 2}$) varying with displacement x_1, (**f**) G-function ($g_{3,\ 4}$) varying with displacement x_1, (**g**) (x_1, x_2) plane (master1), (**h**) (y_1, y_2) plane (master2), (**i**) (w_1, w_2) plane (slave1). (Control parameters: $k_1 = 5$, $k_2 = 20$)

the nonsynchronization of the three state variables x_2, y_2 and w_2 does not possess the G-functions $g_3 < 0$ and $g_4 > 0$. In Fig. 9.4e, f, all G-functions changing with the displacement x_1 are also presented to meet the conditions in Eqs. (9.35–9.38), which are in agreement with the time histories of G-functions in Fig. 9.4d. For example, in

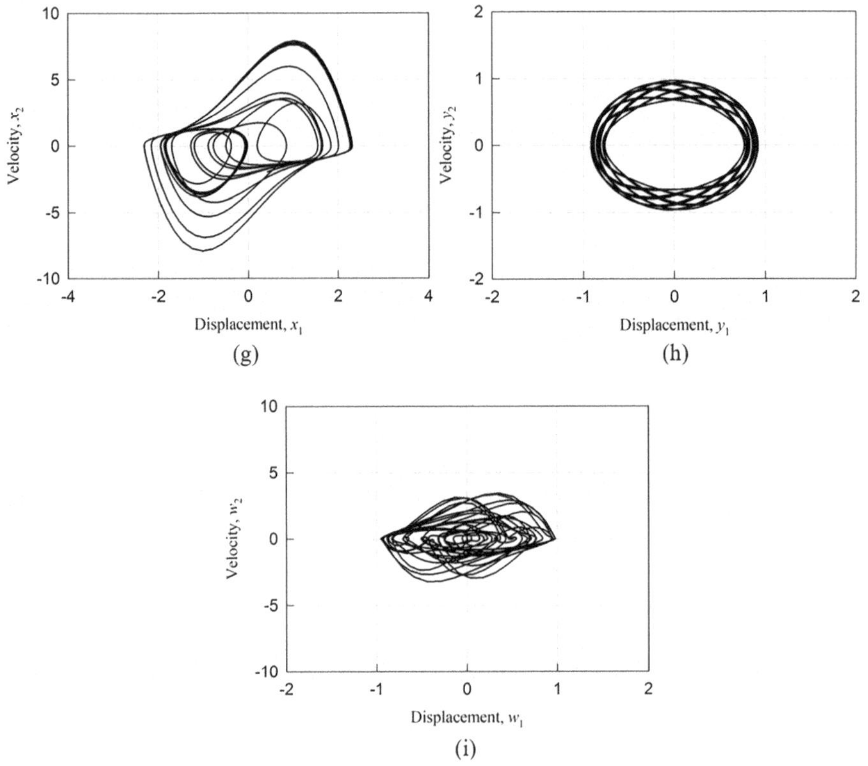

Fig. 9.5 (continued)

Fig. 9.4c, when $t = 0.6990s$, the three system enters the nonsynchronized, and the red line is separated from the zero curve. In Fig. 9.4d, the G-function g_3 crosses the zero point and gradually becomes $g_3 > 0$, which means the velocity of Duffing system cannot obtain the combination synchronization with that of the van der Pol and the pendulum.

In Fig. 9.5a–d, the time histories of displacements, velocities, the error curves, and G-functions for the three systems are depicted, respectively. All the full synchronizations under sinusoidal constraints are labeled with gray areas. In Fig. 9.5c, the error curves of synchronizations are equal to zero with the range of time. The G-functions varying with state variable x_1 are also presented in Fig. 9.5e, f. From the synchronization conditions in Eqs. (9.37 and 9.38), the G-functions $g_i (i = 1, 2)$ for the state variable w_1 synchronized with $\sin(p_1 x_1 + p_2 y_1)$ always exist $g_1 < 0$ and $g_2 > 0$, and the G-functions $g_i (i = 3, 4)$ for the state variable w_2 synchronized with $(p_1 x_2 + p_2 y_2) \cos (p_1 x_1 + p_2 y_1)$ are always satisfied with $g_3 < 0$ and $g_4 > 0$. The phase trajectories of the two master systems and the controlled slave system are plotted in Fig. 9.5g–i, where the full combination synchronization can be clearly observed.

References

Luo, A. C. J. (2008). A theory for flow switchability in discontinuous dynamical systems. *Nonlinear Analysis: Hybrid Systems, 2*(4), 1030–1061.

Luo, A. C. J. (2009a). *Discontinuous dynamical systems on time-varying domains*. HEP-Springer.

Luo, A. C. J. (2009b). A theory for synchronization of dynamical systems. *Communications in Nonlinear Science and Numerical Simulation, 14*, 1901–1951.

Luo, A. C. J. (2012). *Discontinuous dynamical systems*. Higher Education Press/Springer.

Min, F. H., & Ma, H. Y. (2020). The mechanism of switching combination synchronization for three distinct nonautonomous systems under sinusoidal constraints. *Nonlinear Dynamics, 100*(1), 475–492.

Index

A
Accessible sub-domain, 9, 10

B
Bifurcation scenarios, 29–31, 45–49, 65–68

C
Chua circuits, 113, 114, 122, 126, 128, 130, 135, 140–143, 145, 148

D
Discontinuous dynamical systems, 1–5, 9, 13–17, 19–21, 23–25, 57, 77, 83, 84, 99, 113, 122, 126, 128, 151
Duffing like oscillator, 4, 23–34
Duffing oscillator, 3, 5, 151–153, 163

E
Edge interaction theory, 88–90

F
Field programmable gate array (FPGA), 4, 5, 73, 74
FitzHugh-Nagumo neuron, 37
Full synchronizations, 3, 107–109, 111, 112, 134–136, 148, 164, 168, 169

G
G-function
 *k*th-order-*G*-functions, 12
Grazing bifurcation, 45–48
Gyroscope systems, 5, 93–99, 101–105, 107–112

H
Higher-order G-functions, 11
Hindmarsh-Rose neuron, 4, 57–74

I
Inaccessible sub-domain, 9
Instantaneous interaction, 85–89
Interaction constraints
 *j*th non-connected interaction, 78
Invariant domains, 3, 5, 111, 113, 131, 138, 141, 142

M
Mapping structures, 4, 27–29, 31–33, 44–46, 48–52, 64–68

N
Non-passable
 $(2k_i : 2k_j)$ non-passable flow of the first kind, 16

F. Min, *Discontinuous Dynamics and System Synchronization*, Nonlinear Systems and Complexity 39, https://doi.org/10.1007/978-3-031-66648-3

Non-passable (*cont.*)
$(m_i : m_j)$ non-passable flow of the second kind, 18

P
Parameter maps, 5, 31, 32, 51, 52, 57, 93, 109, 113, 134, 136–137, 151, 163–164, 166
Partial synchronizations, 93, 107–110, 135–137, 164, 167
Pendulums, v, 3, 5, 151, 152, 163–165, 169
Period-doubling bifurcation, 29, 31, 45–48, 66, 67
Pitchfork bifurcation, 29, 31, 45, 47, 48

R
Resultant dynamical system, 83

S
Saddle-node bifurcation, 45–48, 66, 67
Semi-passable
$(2k_i : m_j)$ semi-passable, 14, 15
Signum function, 4, 5, 23, 24
Singular edge, 88
Sink flow
$(2k_i : 2k_j)$-sink flow, 16
Source flows, 2, 4, 17, 85, 99
Source interactions, 86, 87
State variables, 24, 40, 52–56, 60, 80, 82, 94, 114, 137, 139–142, 145, 164, 168, 169
Stick interaction, 85, 86
Switching function dynamics, 4, 5, 37–56, 161
Switching sets, 4, 27–29, 40, 43, 44, 52, 62, 64–65
Synchronization dynamics, 93, 126–142, 155–159
System interaction theory, 77–90

T
Tangential flow
$(2k_\alpha - 1)$th-order tangential flow, 20
Total derivative operators, 11

V
van der Pol system, 2, 5, 151, 152, 163–165, 169
Vector fields, 9, 10, 12, 24–26, 41, 42, 60, 62, 63, 78, 79, 81, 82, 94, 95, 115, 118, 124, 152, 159

The manufacturer's authorised representative in the EU is Springer Nature Customer Service Centre GmbH, Europaplatz 3, 69115 Heidelberg, Germany. If you have any concerns regarding our products, please contact ProductSafety@springernature.com

Printed and bound by CPI Group (UK) Ltd, Croydon, CR0 4YY

01/10/2026

02225987-0007